Copyright

Please always consult a trained professional before making any decision regarding treatment of yourself or others.

For more information, email membele1@outlook.com.

Table of Contents

Copyright 2

Table of contents 4

Preview: On the Origins and Evolution of This Work 9

Chapter 1 11

1.1. A beginning of Irregular Sequences and Functions 11

1.1.1. Unearthing Irregular Sequences 11

1.1.2. Determining Control Function Values from a Known Term 18

1.1.3. Deriving n from Known Control Function Values 19

1.2. On the Nature and Necessity of Control Functions in Irregular Sequences 22

1.2.1. The role of Repeats in Control Functions 23

1.3. Constructing an Irregular Sequence via Alternating Star Structures 28

1.3.1. Alternating Star Structure Sequence 28

1.3.2. Construction and Analysis of an Irregular Alternating Sequence 34

1.3.3. Derivation and Analysis of an Irregular Numerical Sequence Using Nested Summations 41

1.4. Alternative Forms of Irregular Sequences 47
1.4.1. Formulation of Another Distinct Sequence 47
1.4.2. Analysis and Representation of another Numerical Sequence 51

Chapter 2 56
2.1. The Addition Interchanges Process 56
2.1.1. The First Addition Interchanges Process 56
2.1.2. A Second Addition Interchanges Process 60
2.2. The Summation of Common Differences 63
2.2.1. Analysis of a Non-Uniform Sequence via Summation of Common Differences 63
2.2.2. Nested Summation and Sequence Construction 66

Chapter 3 70
3.1. Irregular Functions and Control-Mediated Mapping 70
3.1.1. The Role of Gradients in Irregular Functions 73
3.1.2. Bijectivity of Irregular Functions 76
3.1.3. The Summation Function as a Functional Relation 78
3.2. Composite Linear Functions and Graphical Structure 80
3.3. The Other Part of the Theory of Straight Lines 81

Chapter 4 109
4.1. The Derivatives 109
4.1.1. Partial Derivatives of Irregular Functions 109
4.1.2. Derivatives of a Summation-Based Irregular Function 115
4.1.3. Partial Derivatives and Gradient Analysis of a Composite Function 126
4.2. Derivatives and Limits in the Context of Discrete and Continuous Functions 129
4.2.1. Analysis of Gradients 129

Chapter 5 144
5.1. Quadratic Nature of Summation Functions 144
5.1.1. Analysis of the First Summation Function 144
5.1.2. Analysis of the Second Summation Function 147
5.2. Irregular Second-Degree Functions 150
5.2.1. The Curvy-Straight Function with Diminishing $g(x)$ Term 150
5.2.2. Geometric Interpretation: The Shape of the Function 156
5.3. Analysis of a Curvy-Straight Chain Function with Non-Diminishing Linear Components 161
5.4. Exploring the Curvy Function 169

Chapter 6 178

6.1. Transformations of Irregular Functions 178
6.2. Other Transformations 182
6.2.1. The Two-Gradient Irregular Function 182
6.2.2. Addition-Based Transformations of Sigma Functions 186
6.2.3. Double Sigma Addition Functions 193

Chapter 7 198

7.1. Application Examples in Commerce 198
7.1.1. Application Example: Cake Packing and Cost Modelling 198
7.1.2. Modelling the Coffee-Making Process Using the Equation $y = df(x) \equiv 2f(x)$ 201
7.1.3. Cost Modelling in Factory Production Using Control Functions 209
7.1.4. Modelling Business Profit and Investment Using a Summation Function 212
7.2. Application Examples in Science 216
7.2.1. Application of the Two Common Differences Equation in Fluid Mechanics 216
7.2.2. Modelling a Three-Liquid Dispensing System Using Control Functions 221

7.2.3. Modelling Motion with Irregular Sequences and Functions 224
7.2.4. Irregular Sequence Control Response Model Compared to PID Control Response Model 230

Chapter 8 240

8.1. A Further Investigation 240
8.2. Function Behaviour Between Perpendicular Lines 243
8.3. Comparing Regular and Irregular Linear Functions 247
8.4. Constructing the ninth Line Segment 250

Conclusion 253

References 254

Preface: On the Origins and Evolution of This Work

This book represents the culmination of several years of research, and iterative refinement in the field of mathematics—particularly in the introduction of irregular sequences, control functions, and real-world operational frameworks. Many of the foundational ideas presented herein were first explored in a series of preprint articles, which served as early platforms for introducing new sequences/functions and their pedagogical strategies.

These preprints, shared openly with the academic community, allowed for valuable feedback and conceptual evolution which is still ongoing. While the core mathematical insights remain intact, the material in this book has been significantly expanded with application examples, restructured, and recontextualized to serve a broader educational purpose. Readers will find new comparative analyses, schematic diagrams, and real-world modelling examples that were not present in the original drafts.

Where appropriate, citations to the original preprints are provided to acknowledge their role in the development of this work. However, the book should be viewed as a distinct and comprehensive treatment—designed not only to advance mathematical constructs but also to make them accessible, transparent, and pedagogically robust for professionals and advanced students alike.

There has been an attempt to work out the second degree irregular functions which tend to be chain-like functions. This chain-like behaviour is first seen displaying in summation functions that map different linear functions as one function or many straight lines as one characteristic graph. There has also been an endeavour to include some applications of some of the discussed irregular sequences and functions.

I wrote this book because I felt we are going a lot forward and advanced with regular mathematical objects without having studied the irregular counterparts in their entirety. Therefore my focus when writing this book was that we try and balance the developments of both especially that we begin the study of the irregular mathematical object which might help us to facilitate or obtain solutions to some old and new problems in different fields of study (maybe tackle old problems in a different setting).

Chapter 1

Mathematics is often framed by predictability: patterns that follow uniform rules, functions with clear behaviors, sequences marked by regular spacing. Yet beyond this familiar terrain lies an intriguing frontier—**irregular sequences**—where the rules are richer, less obvious, and demand new methods of inquiry. This chapter marks our departure from traditional linear progressions and enters a setting where **multiple common differences** shape the behaviour of a sequence.

1.1. A beginning of Irregular Sequences and Functions

1.1.1. Unearthing Irregular Sequences

Irregular sequences often arise when enumerating points on the vertices of regular two-dimensional shapes—such as triangles. Table 1.1.1.1 illustrates vertex-based sequences, where Vertex 1 and Vertex 3 exhibit irregular patterns due to multiple alternating common differences. In contrast, Vertex 2 presents a standard arithmetic progression characterized by a single constant difference.

This work focuses specifically on constructing and analysing sequences governed by two or more recurring common differences. In Table 1.1.1.1 below, the sequences are obtained by counting three numbers clockwise on a triangle and then three numbers anti-clockwise. The counting goes on clockwise and anti-clockwise to infinity and as said, it produces sequences that are both regular and irregular on the vertexes of any regular shape (here a triangle).

Table 1.1.1.1: Vertex-Based Sequences

Vertex 1	Vertex 2	Vertex 3
1	2	3
6	5	4
7	8	9
12	11	10
13	14	15
18	17	16

Let us examine the sequence located at Vertex 1:

$$\{a_n\} = 1, 6, 7, 12, 13, 18, 19, 24, \ldots$$

This sequence demonstrates two interleaved common differences, denoted d_1 and d_2:

- $d_1 = T_2 - T_1 = 6 - 1 = 5$
- $d_2 = T_3 - T_2 = 7 - 6 = 1$

These differences persist in an alternating structure throughout the sequence—beginning with d_1, then followed by d_2, cyclically. To formulate a general expression for the nth term T_n, we adopt a method analogous to that used in conventional arithmetic sequences, albeit adapted to accommodate the alternating nature of the differences.

Table 1.1.1.2: Recursive Structure of Terms

$\boldsymbol{T_1 = a}$
$\boldsymbol{T_2 = a + d_1}$
$\boldsymbol{T_3 = a + d_1 + d_2}$
$\boldsymbol{T_4 = a + 2d_1 + d_2}$
$\boldsymbol{T_5 = a + 2d_1 + 2d_2}$
$\boldsymbol{T_6 = a + 3d_1 + 2d_2}$
$\boldsymbol{T_7 = a + 3d_1 + 3d_2}$

Substituting $a = 1$, $d_1 = 5$, and $d_2 = 1$, we validate the accuracy of the model by calculating, for instance:

$$T_7 = 1 + 3(5) + 3(1) = 19$$

To systematize the coefficients of d_1 and d_2 across terms, we introduce control functions $f(n)$ and $g(n)$, which track their respective multipliers.

n	1	2	3	4	5	6	7	8
$f(n)$	0	1	1	2	2	3	3	4
$g(n)$	0	0	1	1	2	2	3	3
T_n	1	6	7	12	13	18	19	24

Table 1.1.1.3: Control Functions and Term Generation

As shown, the coefficients from Table 1.1.1.2 align with the values of $f(n)$ and $g(n)$, confirming their role in capturing the structural rhythm of the sequence. These functions thus serve as control mechanisms, enabling formulation of a general expression for irregular sequences:

$$T_n = \mathrm{a} + d_1 f(n) + d_2 g(n) \qquad (eq.\ 1.1)$$

When substituting the known values (common differences) into Equation (1.1):

$$T_n = 1 + 5f(n) + g(n) \qquad (eq.\ 1.2)$$

Finally, consistency among the control functions and their alignment with n is maintained under the inequality:

$$n > f(n) \geq g(n)$$

This ensures coherent generation of term values, where $f(n)$ and $g(n)$ guide the respective contributions of d_1

and d_2. The resultant sequence may also be represented graphically (see Graph 1.1.1.1) to visualize the trajectory of values over n.

Graph 1.1.1.1: Visualization of Trajectory of Values over n

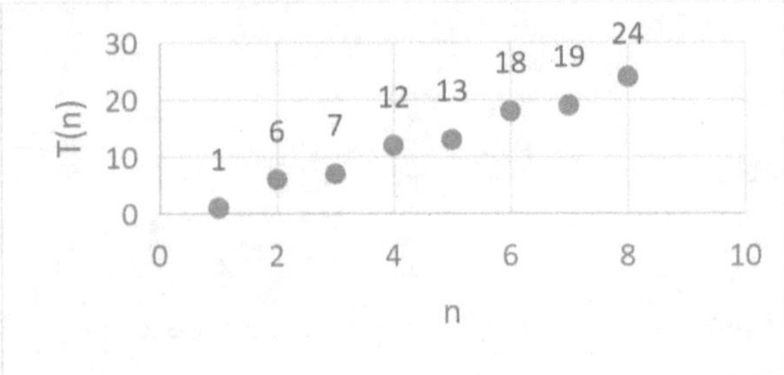

The universal formula for the nth term of irregular sequences like these is

$$T_n = d_1 f(n) \pm d_2 g(n) \pm a$$

We adopt the convention of designating a (the initial term) and d (the common difference) similar to standard sequences. The formula for T_n can serve as the formula for the sequence $\{a_n\}$, for example, $\{a_n\} = 5f(n) + g(n) + 1$. Examining the series of d's in the pattern above (considering only d_1 and not both d_1 and d_2), it creates a sub-sequence that is similarly irregular (refer to table 1.1.1.2). This pattern captures our attention specifically because it relates to the topics we explore in this book. Consider a related (similar) pattern that is like this.

$$T_n = 2, 2, 5, 5, 8, 8, 11, 11, \ldots$$

In this sequence, there is a single common difference d, calculated as $d = T_3 - T_2 = 5 - 2 = 3$. At this stage, we ought to disregard the $d = 0$ in this pattern as if it were non-existent. At this point, it is evident that this $d = 0$ is shown by the control function's capacity to replicate values, even if at different times it indicates a repeated d. This explains how to recognize this zero, which can be understood as the terms recurring. The cardinality of n ($|n|$) indicates the number of repetitions present. Our $a = 2$ (a is the initial term) and $d = 3$ and using this information as we did with the initial sequence, we can derive an equation for the present sequence, which is:

$$T_n = 3f(n) + 2$$

Where $|n| = 2$ and the alignment inequality is $f(n) < n$. (Remember the common difference multiply the control function in the equation and others). *Table 1.1.1.4* below is the table of values for this current function.

Table 1.1.1.4 below displays a sequence exhibiting a repeating-value structure governed by a control function $f(n)$:

Table 1.1.1.4: Exhibiting the Repeating-Value Structure

n	1	2	3	4	5	6	7
$f(n)$	0	0	1	1	2	2	3
T_n	2	2	5	5	8	8	11

The graph corresponding to this sequence is presented in Graph 1.1.1.2.

Graph 1.1.1.2: Visuals of $T_n = 3f(n) + 2$

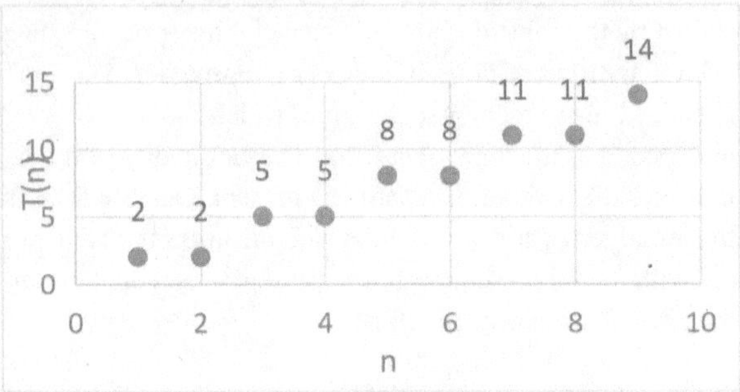

We define the general formulation for such sequences as:

$$T_n = df(n) \pm a$$

Here, d denotes the common difference and a the initial term. Importantly, d may assume positive or negative values—enabling modeling of both increasing and decreasing trends within the sequence.

This structure naturally extends from the behaviour illustrated in the sequence and is reflected in the graphical representation provided in Graph 1.1.1.2. By encoding repetition through the control function $f(n)$, the model accounts for consistent duplications and incremental.

Graph 1.1.1.3: Representation of $T_n = 1 + 5f(n) + g(n)$ as a Function

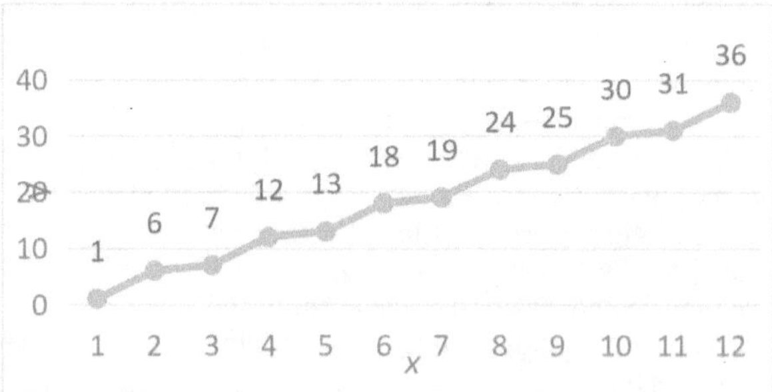

Upon reviewing the irregular sequence:

$$\{a_n\} = 1, 6, 7, 12, 13, 18, 19, 24, \ldots$$

modelled by the general expression:

$$T_n = 5f(n) + g(n) + 1$$

We observe that the differences between selected terms (e.g., 7 - 6 = 1, 12 - 6 = 6, 13 - 6 = 7) are themselves part of the original sequence. This reinforces the internal recursive structure of the sequence, wherein each term can be generated from its predecessor two steps prior:

$$T_n = T_{n-2} + (d_1 + d_2)$$

Here, d_1 and d_2 are the alternating common differences, and their sum $d_1 + d_2 = S_d$ serves as the recursive increment. Rearranging, the recursive relation may also be expressed as:

$$T_n = T_{n-2} + S_d$$

This formulation enables computation of any term T_n given T_{n-2} and the combined difference S_d, offering a systematic way to encode term generation.

Notably, the sequence observed at Vertex 3 shares an identical structural pattern with that of Vertex 1. By applying the general expression $T_n = d_1 f(n) + d_2 g(n) + a$, and substituting the appropriate initial term and positive common differences specific to Vertex 3 pattern, an analogous functional equation can be derived. This parallelism underscores the universality and transferability of the control function model across symmetrical geometric arrangements.

1.1.2. Determining Control Function Values from a Known Term

In the irregular sequence modelled by:

$$T_n = 5f(n) + g(n) + 1$$

We aim to identify the control function values $f(n)$ and $g(n)$ corresponding to a known term—specifically when $T_n = 19$. Given the cardinality of n is 3, the structure of both control functions follows a cyclical progression, as illustrated in Table 1.1.2.1.

Table 1.1.2.1: Determining Term Values of the Current Sequence

n	1	2	3	4	5	6	7	8	9	10	11	12
f(n)	0	1	1	1	2	2	2	3	3	3	4	4
g(n)	0	0	0	1	1	1	2	2	2	3	3	3
T(n)	1	6	6	7	12	12	13	18	18	19	24	24

We begin by setting up and simplifying the equation for $T_n = 19$:

$$19 = 5f(n) + g(n) + 1$$

therefore

$$18 = 5f(n) + g(n)$$

Knowing that the alternating common differences are $d_1 = 5$ and $d_2 = 1$, we consider their sum:

$$S_d = d_1 + d_2 = 6$$

(please note that there is a third common difference we ignored, because it is equal to 0). Dividing 18 proportionally according to the relative weights of d_1 and d_2:

- Contribution from $f(n)$:

$$5/_6 \times 18 = 15$$

- Contribution from $g(n)$:

$$1/_6 \times 18 = 3$$

Due to $f(n) > g(n)$, we accept the smallest viz. 3 of these values because it makes sense that the term 19 cannot be at greater control function values than that. Then $g(n) = 3$ and we substitute it to find $f(n) = 3$.

1.1.3. Deriving n from Known Control Function Values

Definition 1:
Given the cardinality $|n|$ and a constant C, the relationship

between the position n and its corresponding control function value $f(n)$ is governed by:

$$f(n) = \frac{n}{|n|} \pm C$$

This formulation enables bidirectional computation: knowing either $f(n)$ or n, one may recover the other.

Worked Example: Finding n, when $f(n) = 4$

To identify all valid values of n that yield $f(n) = 4$, we first examine the structure of $f(n)$, which follows a repeating pattern based on the cardinality, $|n| = 3$:

$$f(n) = 0, 1, 1, 1, 2, 2, 2, 3, \ldots$$

The three instances of $f(n) = 1$ occur at $n = 2, 3, 4$ (positions) confirming the cardinality constraint. From this, the constant is determined:

- $C = \frac{1}{|n|} = \frac{1}{3}$
- Thus, the complementary form is $1 - C = \frac{2}{3}$

Now examine the fractions of $\frac{n}{|n|}$ for the above positions:

n	$\frac{n}{\lvert n \rvert}$	Adjustment to get $f(n) = 1$
2	$\frac{2}{3}$	Add $\frac{1}{3}$
3	$\frac{3}{3}$	No adjustment
4	$\frac{4}{3}$	Subtract $\frac{1}{3}$

Hence, all three yield $f(n) = 1$ through $\frac{n}{|n|} \pm C$. We now apply this structure to solve for positions yielding $f(n) = 4$:

Case 1:

Assume $f(n) = \frac{n}{|n|} \pm C = \frac{n}{3} + \frac{1}{3}$

$4 = \frac{n_1}{3} + \frac{1}{3}$ Which implies $\frac{n_1}{3} = \frac{11}{3}$ that will give us $n_1 = 11$

Case 2:

Assume $f(n) = \frac{n}{3}$

$4 = \frac{n_2}{3}$ This implies $n_2 = 12$

Case 3:

Assume $f(n) = \frac{n}{3} - \frac{1}{3}$

$4 = \frac{n_3}{3} - \frac{1}{3}$ which implies $\frac{n_3}{3} = \frac{13}{3}$ that will give us $n_3 = 13$

Thus, the three positions at which $f(n) = 4$ occur are:

$$n = 11, 12, 13$$

The selection of a specific value depends on additional context, such as the corresponding value of $g(n)$ or the desired sequence term T_n.

1.2. On the Nature and Necessity of Control Functions in Irregular Sequences

Control functions arise as a formal mechanism for managing structural repetition within algebraically defined irregular sequences. While originally inspired by the concatenation function in computer science—used for string construction and replication—control functions have been reformulated to serve a more algebraic role, specifically enabling substitution patterns within non-standard sequence definitions.

If n is the independent variable, then a control function such as $f(n)$ allows selective substitution of values in place of n, thereby regulating term generation in cases where repetition or alternation is intrinsic to the sequence. This design is essential when ordinary arithmetic formulations fall short due to irregular progression.

A pivotal concept in this framework is the **cardinality** $|n|$, which—though traditionally defined as the number of elements in a set—is repurposed here to denote the number of successive positions in which the control function $f(n)$ retains a constant value. That is, $|n|$ quantifies the length of uniform segments governed by $f(n)$, acting as a repetition counter for values of n over which the output remains unchanged.

Consider a mapping:

$$f\{n_1, n_2\} \rightarrow 2,2$$

Under this paradigm, the image is treated as a **set of algebraic values** rather than a **string of digits**. This distinction is crucial: while string concatenation would interpret $\{2, 2\}$ as "22," algebraic operations demand set-theoretic treatment where repetition does not alter identity—i.e., $\{2, 2\}$ simplifies to $\{2\}$. Thus, we assert:

- $\{2, 2\} = 2$, not "22"
- $\{a, a\} = a$, implying both $(f\{n_1\} = a)$ and $(f\{n_2\} = a)$

Such structural consistency ensures that control functions are suitable for algebraic manipulation and integration into sequence-generating formulas.

Ultimately, control functions are indispensable for modelling and analysing irregular sequences. The inherent repetition of values, once a source of analytical complexity, is now central to their formulation. By leveraging ($f(n)$) and ($|n|$), we gain a systematic method to encode and explore these patterns, transforming repetition from a problem into a powerful modelling principle.

1.2.1. The Role of Repeats in Control Functions

In this sub-section, we explore scenarios in which the functions $f(n)$ or $g(n)$ exhibit repetition—or lack thereof—in their output values. This distinction will help us formalize the existence and behaviour of control functions within the broader context of this work.

Case 1: Absence of Repeats

When repetition is absent, the mapping $f\{n\} \to X$, with $X = \{a\}$, implies a cardinality $|n| = 1$. This means each input n corresponds to a unique value, even if sets like $\{a, a\}$ collapse to $\{a\}$. In such cases, $f(n) = n$, as illustrated in Table 1.2.1.1. The sequence thus becomes regular, with the identity alignment $f(n) = n$ and $n \in \mathbb{N}$.

We demonstrate this using the equation:

$$T_n = 3f(n) + 2$$

Since $f(n) = n$, this simplifies to

$$T_n = 3n + 2$$

, yielding:

Table 1.2.1.1. A Regular Sequence where $T_n = 3n + 2$ for $n \geq 1$.

n	1	2	3	4	5	6	7
$f(n)$	1	2	3	4	5	6	7
T_n	5	8	11	14	17	20	23

This substitution introduces no contradictions, illustrating a straightforward case of a control function maintaining regularity.

Case 2: Still No Repeats, but $f(n) \neq n$.

Here, the cardinality remains $|n| = 1$, but we now allow the inequality $f(n) < n$ (or $f(n) > n$) as the basis for alignment. We again apply the equation:

$$T_n = 3f(n) + 2$$

Suppose $f(n) \geq 0$ and $n \geq 1$, with $n \in \mathbb{N}$. This yields a different starting value:

Table 1.2.1.2. An example where Alignment is Non-Identity: $f(n) \neq n$.

n	1	2	3	4	5	6	7
$f(n)$	0	1	2	3	4	5	6
T_n	2	5	8	11	14	17	20

In such cases, knowledge of specific $f(n)$ values is essential to define the sequence.

On the Meaning of Cardinality and Zero Repeats

What does $|n| = 0$ represent? It denotes zero repetitions, not the absence of n itself. This can align with the concept of a single occurrence per n, akin to the factorial identity 0! = 1! = 1.

If $|n| = \emptyset$ and $n = \emptyset$, we interpret this as a complete absence of inputs—no n exists—rendering the definition of sequences or functions void unless extended into complex domains. However, if $n \neq \emptyset$, the situation is analogous to $|n| = 0$, indicating no repeats but active sequence participation.

In summary, zero repetition is associated with $|n| = 0$ or $|n| = 1$, and true repetition emerges only when $n \geq 2$.

Case 3: Repeating Values and Irregularity in Sequences

When the cardinality of the set n is a constant $C \geq 2$, we encounter repeated values. Specifically, if $|n| = C$, then $f\{n_1, n_2\} \to a$, a meaning the value of $f(n)$ repeats C times. For instance, when $C = 2$, we obtain pairs such as $f\{n_1, n_2\} \to a$, leading to outputs to $T_n = 3f(n) + 2$, as illustrated in Table 1.2.1.3 below.

n	1	2	3	4	5	6	7
$f(n)$	0	0	1	1	2	2	3
T_n	2	2	5	5	8	8	11

Table 1.2.1.3

As shown above, repetitions in $f(n)$ induce irregularities in the output sequence T_n. Such repetition can manifest not only as identical value pairs (e.g. 5, 5) but also as structural cycles—such as odd or even sequences like 1, 3, 5, 7 or 2, 4, 6, 8—where values differ but follow a repeating pattern.

Case 4: Infinite Repeats and Constant-Valued Functions

Now consider the case of infinite repetition, where the cardinality of n approaches infinity, i.e. $|n| = \infty$. In this scenario,$n = \{n_1, n_2, ...\}$, and $f(n)$ remains constant across all values. Consequently, $T_n = 3f(n) + 2$ yields a uniform sequence, akin to the constant function $h(x) = 8$ in regular functional analysis. This condition corresponds to functions of the form $y = C$, where C is constant.

Table 1.2.1.4: Representation of a Function that has an Infinite Repetition

n	1	2	3	4	5	6	7
$f(n)$	2	2	2	2	2	2	2
T_n	8	8	8	8	8	8	8

This represents the limit case $f\{n_1, n_2, ...\} \to a, a, ...$, a hallmark of constant-valued irregular functions.

Determining Repetition Length and Cycle Structure

The number of repeats, $|n|$, can be inferred through:

- Direct observation from the sequence or its tabular representation.
- Knowledge of $|n|$ from structural definitions.
- Analysis of initial values in $f(n)$.

Cycle Structure and Cardinality in Irregular Sequences

The **cardinality of n**—denoted $|n|$—serves as a measure of the cycle lengths within a given sequence and indicates how frequently $f(n)$ repeats a value. These cycles may exhibit one of the following forms:

- **Constant values** (e.g., repeated identical entries),
- **Monotonically increasing values** (e.g., 0, 1, 2, 3),
- **Monotonically decreasing values** (e.g., 5, 4, 3, 2).

Consider the sequence$\{a_n\} =$ $0, 1, 2, 3, 5, 6, 7, 8, 10, 11, 12, 13$, governed by the function:

$$T_n = (n - 1) + f(n)$$

Where $f(n)$ takes in the set $\{0, 0, 0, 0, 1, 1, 1, 1, 2, 2, 2, 2, ...\}$ as it's values.

This sequence comprises distinct **cycles of length four**, such as:

- **Cycle 1**: 0, 1, 2, 3,
- **Cycle 2**: 5, 6, 7, 8,
- **Cycle 3**: 10, 11, 12, 13.

Each cycle is **internally regular**, with a common difference $d = 1$. However, the transition from one cycle to the next introduces a jump of $d = 2$, contributing to a **compound irregularity**. Thus, while individual cycles maintain a predictable structure, their aggregation into a unified sequence results in an overarching irregular pattern.

A Dual-Level Investigation

This framework necessitates a **two-tiered analysis**:

1. The **regular behaviour within each cycle**,
2. The **irregular composition across cycles**.

To address this complexity, we examine the nature and structure of such **irregular sequences or functions**. Toward the end of Section 1.3, a more advanced construct emerges: a function resembling a **vector-valued mapping**, potentially offering a multidimensional approach to modelling cyclical behaviour.

This extension—while promising—will be reserved for a dedicated investigation. For now, let us delve into examples that illustrate the interaction between cardinality, cycle type, and overall sequence behaviour.

1.3. Constructing an Irregular Sequence via Alternating Star Structures

1.3.1. Alternating star structure sequence

To build an irregular numeric pattern, we begin with a visual and symbolic model: two overlapping four-bladed stars—one facing north, one northeast (the star facing north has the other blades also facing East, South and West and the one facing northeast has blades also facing southeast, southwest and northwest). Each star features a distinctive dot indicating the starting point for traversal. Movement proceeds counter-clockwise, beginning with the dot on the star with blades that start facing north.

On the star with the first blade facing north, we count the first four **odd numbers**—namely 1, 3, 5, 7—one per blade, advancing with a **common difference of 2**. This completes one cycle. Next, we **jump** to the star with the first blade facing northeast, starting at its own dot, and count the first four **even numbers**, 8, 10, 12, 14, also spaced by a common difference of 2.

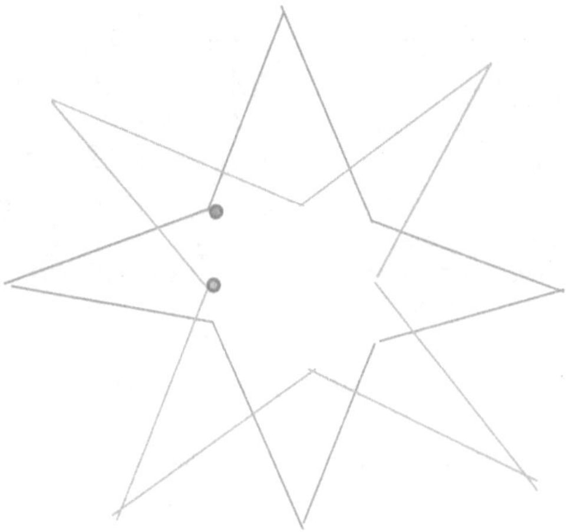

The **jump between stars** introduces a secondary difference: a constant increment of **1** that links the terminal value of one cycle to the initial value of the next.

This alternating process of traversing blades and shifting between stars generates a composite sequence:$\{a_n\} = 1, 3, 5, 7, 8, 10, 12, 14, 15, 17, 19, 21, 22, ...$

Each cycle has:

- An **internal difference**, $d_1 = 2$, used within blades of the same star.
- A **transitional difference**, $d_2 = 1$, used when jumping between stars.

Recursive Structure and Closed-Form Expression

Let the initial term be $a = 1$. Then the sequence $\{T_n\}$ can be expressed with layered increments:

$T_1 = a$
$T_2 = a + d_1$
$T_3 = a + 2d_1$
$T_4 = a + 3d_1$
$T_5 = a + 3d_1 + d_2$
$T_6 = a + 4d_1 + d_2$
$T_7 = a + 5d_1 + d_2$
$T_8 = a + 6d_1 + d_2$
$T_9 = a + 6d_1 + 2d_2, etc$

Table 1.3.1.1

From this, we define the general term using two auxiliary functions:

- $f(n)$: counts the number of transitions between stars (e.g., from odd to even),
- $g(n)$: captures the accumulation of d_2 due to those transitions.

Thus, the closed-form expression becomes:

$$T_n = a + [(n-1) - f(n)]d_1 + g(n)d_2$$

This formulation allows us to preserve the irregular nature while maintaining computational tractability. The term $[(n-1) - f(n)]$ in front of d_1 gives the pattern of the first common difference, d_1, e.g., the $d\ for\ T_7$ is $[(7-1) - 1]d_1 = 5d_1$ and the lone control function viz. $g(n)$ gives the numbers in front of d_2.

Analysis of Control Functions and Irregular Sequences

We now examine the control functions $f(n)$ and $g(n)$, as presented in **Table 1.3.1.2** below:

Table 1.3.1.2: Visualising the Control Functions

n	1	2	3	4	5	6	7	8	9
$f(n)$	0	0	0	0	1	1	1	1	2
$g(n)$	0	0	0	0	1	1	1	1	2
T_n	1	3	5	7	8	10	12	14	15

From the table, it is evident that the control functions $f(n)$ and $g(n)$ are equal for all values of n, and satisfy the condition $f(n) = g(n) < n$. Consequently, either control function value may be used interchangeably in the general equation. It is also essential to assign distinct function variables to the control functions when dealing with different common differences. Furthermore, the cardinality of n in this sequence is 4, denoted as $|n| = 4$.

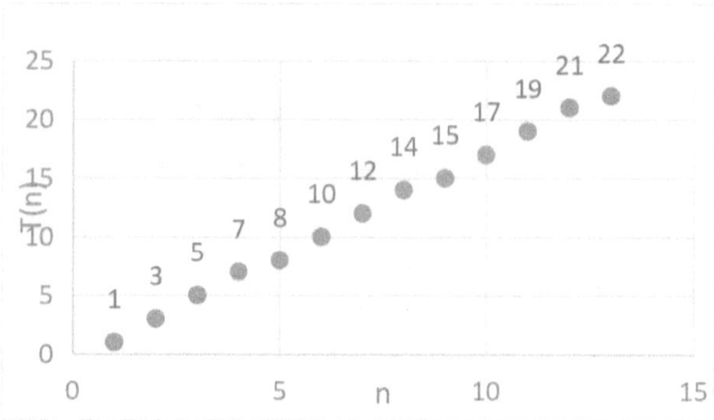

Graph 1.3.1.1: Constructing Irregular Patterns

Irregular patterns can be constructed using similar methodologies or alternative valid approaches. For instance, one may define the pattern of d_2 to differ from that of d_1, or vice versa. As demonstrated, the incorporation of control functions facilitates the analysis and simplification of irregular sequences and their associated functions.

Determining n in the Sequence Equation

To determine the value of n within the given sequence, consider the following parameters:

- $T_n = 14$
- $a = 1$
- $f(n) = 1$
- $g(n) = 1$
- $d_1 = 2$
- $d_2 = 1$

Using the sequence formula:

$$T_n = a + [(n-1) - f(n)]2 + g(n)$$

Substituting the known values:

$$14 = 1 + [(n-1) - 1]2 + 1$$

Simplifying:

$$14 = 2n - 4 + 2$$

$$2n = 16$$

$$then\, n = 8$$

This confirms that $n = 8$ satisfies the equation. The corresponding graph of the sequence is illustrated in **Graph 1.3.1.1** (above).

Cyclical Structure and Sub-Patterns in Irregular Sequences

Further exploration of the sequence reveals a cyclical structure composed of alternating odd and even number groupings. These groupings can be organized into discrete cycles, each containing four consecutive terms. For instance:

- $Cycle\ 1: 1, 3, 5, 7$
- $Cycle\ 2: 8, 10, 12, 14$
- $Cycle\ 3: 15, 17, 19, 21$
- $Cycle\ 4: 22, 24, 26, 28$
- $Cycle\ 5: 29, 31, 33, 35$
- $Cycle\ 6: 36, 38, 40, 42$

This pattern can be generalized using the following vector-valued function:

$$w(x) = \{7x - 6; 7x - 4; 7x - 2; 7x\}$$

Where $x \in \mathbb{N}$ denotes the cycle index. For example, substituting $x = 3$ yields:

$$w(3) = \{15, 17, 19, 21\}$$

These cycles represent a structured system of values governed by linear expressions. The ability to generate each cycle using a single parameterized function facilitates the reconstruction of the original sequence through sequential grouping.

Generalized Sub-Pattern Equation

In addition to the main sequence, a sub-pattern can be described by the generalized form:

$$T_n = [(n \pm 1) \pm f(n)]d$$

This formulation captures the behaviour of sequences with alternating common differences. For example, consider the sequence:

$$a_n = 0, 1, 2, 3, 3, 4, 5, 6, 6, ...$$

Here, the sequence exhibits two distinct common differences:

- $d_1 = 1$
- $d_2 = 0$

Such sub-patterns, embedded within broader irregular sequences, warrant further analytical investigation due to their structured yet non-uniform nature.

1.3.2. Construction and Analysis of an Irregular Alternating Sequence

To initiate the construction of a sequence $\{a_n\}$, we consider two linear functions that are perpendicular to each other:

- $y_1 =- 2x + 1$
- $y_2 = \frac{1}{2}x + 1$

These functions possess gradients $m_1 =- 2$ and $m_2 = \frac{1}{2}$, respectively. The perpendicularity of the lines is evident from the product of their gradients: $m_1 \times m_2 =- 1$. These equations can be written in cyclical structure as in the end of previous sub-section. This is because irregular sequences studied herein are related to the cycles equation looked at in that previous sub-section.

Tabulation of Function Values

The values of the functions for integer inputs $x = 1, 2, ..., 8$ are presented in Table 1.3.2.1:

x	1	2	3	4	5	6	7	8
$-2x + 1$	-1	-3	-5	-7	-9	-11	-13	-15
$\frac{1}{2}x + 1$	$1\frac{1}{2}$	2	$2\frac{1}{2}$	3	$3\frac{1}{2}$	4	$4\frac{1}{2}$	5

Table 1.3.2.1

Formation of the Alternating Sequence

By interleaving the outputs of y_1 and y_2, we obtain the following alternating sequence:

$$\{a_n\} =-1, 1\frac{1}{2}, -3, 2, -5, 2\frac{1}{2}, -7, 3, ...$$

This sequence may be reformulated to highlight its structure:

$$\{a_n\} =-\frac{2}{2}, \frac{3}{2}, -\frac{6}{2}, \frac{4}{2}, -\frac{10}{2}, \frac{5}{2}, -\frac{14}{2}, \frac{6}{2}, ...$$

Analysis of Common Differences

The sequence of common differences $\{d_n\}$ derived from $\{a_n\}$ is itself alternating:

$$\{d_n\} = \frac{5}{2}, -\frac{9}{2}, \frac{10}{2}, -\frac{14}{2}, ...$$

Extracting the numerators and disregarding the alternating signs yields:

$$\{d_{n1}\} = 5, 9, 10, 14, 15, 19, 20, ...$$

This sequence conforms to a generalized form of second-order linear recurrence:

$$T_n = d_1 f(n) \pm d_2 g(n) \pm a$$

Where:

- $a = 5$
- $d_1 = 4$
- $d_2 = 1$

Thus, the governing expression becomes:

$$d_{n1} = 4f(n) + g(n) + 5$$

Definition of Auxiliary Functions

The functions $f(n)$ and $g(n)$ are defined as follows (see table below):

n	1	2	3	4	5	6	7
$f(n)$	0	1	1	2	2	3	3
$g(n)$	0	0	1	1	2	2	3

Table 1.3.2.2

Reconstruction of the Original Sequence

To recover the original sequence $\{a_n\}$, we apply the alternating sum of the common differences to the initial term $a_1 =- 1$.

T_n	Relation of the d with 1st term	Value of T_n
T_1	-1	-1
T_2	$-1+\frac{5}{2}$	$1\frac{1}{2}$
T_3	$-1+\frac{5}{2}-\frac{9}{2}$	-3
T_4	$-1+\frac{5}{2}-\frac{9}{2}+5$	2

Table 1.3.2.3

The closed-form expression is:

$$T_n =- 1 + \frac{1}{2}\sum_{i=1}^{n-1} (-1)^{i-1}(4f(n) + g(n) + 5)$$

Verification Example

To compute T_7, we evaluate:

$$T_7 =- 1 + \frac{1}{2}(5 - 9 + 10 - 14 + 15 - 19) =- 7$$

This result confirms the validity of the derived expression for T_n.

Extension of Irregular Sequence Analysis and Generalizations

Building upon the previously defined alternating sequence derived from two perpendicular linear functions, we now introduce a set of general equations that describe similar irregular patterns. These formulations are based on the structure of the original sequence and its associated sequence of common differences.

General Equations for Irregular Sequences

Let $f(n)$ and $g(n)$ be auxiliary functions as previously defined and let d_1, d_2, b, c be constants. The following equations represent various forms of irregular sequences:

(1.3.1) General Form Including Initial Term

$$a_n = a \pm \frac{1}{c}\sum_{i=1}^{n-1}(-1)^{i-1}(d_1 f(n) \pm d_2 g(n) \pm b)$$

This equation generalizes the structure of the original sequence, incorporating an initial term a and alternating signed increments.

(1.3.2) Common Differences as a Sequence (Excluding Initial Term)

$$a_n = \frac{1}{c}\sum_{i=1}^{n-1}(-1)^{i-1}(d_1 f(n) \pm d_2 g(n) \pm b)$$

This form treats the sequence of common differences as an independent sequence.

(1.3.3) Unscaled Summation Form

$$a_n = \sum_{i=1}^{n-1}(-1)^{i-1}(d_1 f(n) \pm d_2 g(n) \pm b)$$

Here, the scaling factor $\frac{1}{c}$ is removed, yielding a raw summation of alternating terms.

(1.3.4) Term-wise Expression Without Summation

$$a_n = (-1)^{n-1}(d_1 f(n) \pm d_2 g(n) \pm b)$$

This expression provides the value of each term directly, without accumulation.

Note: Each of these equations may be further generalized by multiplying the entire expression by a rational factor $\frac{p}{q}$, thereby generating distinct sequences with similar structural properties.

Revisiting the Original Sequence via Cycles

Let us reconsider the original sequence through the lens of **cyclic pairs**, where each cycle consists of two consecutive terms derived from the functions:

- $y_1 = -2x + 1$
- $y_2 = \frac{1}{2}x + 1$

Cycle Representation

Each cycle is defined as:

$$cycle(x) = (y_{x-1}, y_x) = \left(-2x + 1, \frac{1}{2}x + 1\right)$$

For example:

- $Cycle\ 1: (-1, \frac{3}{2})$
- $Cycle\ 2: (-3, 2)$
- $Cycle\ 3: (-5, \frac{5}{2})$, etc.

Increment Within Cycles

The increment within each cycle is given by:

$$\Delta y = (y_x - y_{x-1}) = (\frac{1}{2}x + 1) - (-2x + 1) = \frac{5}{2}x$$

This increment corresponds to the **Euclidean distance** between the two points in each cycle:

$$D = \sqrt{(x - x)^2 + (\frac{1}{2}x + 1 - (-2x + 1))^2}$$

$$= \sqrt{(\frac{5}{2}x)^2}$$

$$= \frac{5}{2}x$$

Thus, the increment is equivalent to the vertical distance between the two function values at a given x.

Sequence of Cycles

If we consider the second term subtract the first of each cycle as representative, we obtain a new sequence:

$$\{C_n\} = 2\tfrac{1}{2}, 5, 7\tfrac{1}{2}, 10, \ldots$$

This sequence exhibits a regular pattern with a common difference of $\frac{5}{2}$, as shown by:

$5 - \frac{5}{2} = \frac{5}{2}, \frac{15}{2} - 5 = \frac{5}{2}$, etc.

Hence, the sequence of cycles forms a regular arithmetic progression.

Considerations for the next section

In subsequent analysis, we will explore sequences derived from **reciprocal gradients** of linear functions. Specifically, we will examine the pair:

$$w(x) = [2x + 1, \frac{1}{2}x + 1]$$

This will allow us to investigate further structural relationships and extend the family of irregular sequences.

1.3.3. Derivation and Analysis of an Irregular Numerical Sequence Using Nested Summations

As previously referenced, we now proceed to enumerate the point sets corresponding to the given linear functions. These are presented in Table 1.3.3.1 below:

n	**1**	**2**	**3**	**4**	**5**	**6**	**7**	**8**
$2n+1$	3	5	7	9	11	13	15	17
$\frac{1}{2}n+1$	$1\frac{1}{2}$	2	$2\frac{1}{2}$	3	$3\frac{1}{2}$	4	$4\frac{1}{2}$	5

Table 1.3.3.1

From these cycles, we construct an **irregular sequence** defined as:

$$\{a_n\} = 3, 1\frac{1}{2}, 5, 2, 7, 2\frac{1}{2}, 9. \ldots$$

Each point-set is expressed as(n, cycle) = $(n, (y_{n-1}, y_n))$. The sequence is classified as irregular due to the absence of consistent linear, quadratic, or geometric progression in its differences. To simplify, we rewrite the sequence as:

$$\{a_n\} = \frac{6}{2}, \frac{3}{2}, \frac{10}{2}, \frac{4}{2}, \frac{14}{2}, \frac{5}{2}, \frac{18}{2}, \dots$$

Given the uniform denominator, we focus on the **numerator pattern**:

$$\{a_{n1}\} = 6, 3, 10, 4, 14, 5, 18, 6, \dots$$

To derive the general term of this sequence, we analyse its **first-order differences**:

$$\{d_{n1}\} = -3, 7, -6, 10, -9, 13, -12, 16, -15, \dots$$

This alternating irregular pattern prompts a second-order difference analysis:

$$\{d_{n2}\} = 10, -13, 16, -19, 22, -25, 28, -31, \dots$$

By isolating the positive terms, we obtain a regular arithmetic sequence with a common difference d = 3. The general term of this positive sequence is:

$$3n + 7$$

To account for the alternating signs, we incorporate $(-1)^{n-1}$, yielding:

$$d_{n2} = (-1)^{n-1}(3n + 7)$$

We now reconstruct the first-order differences using cumulative summation:

$d_{11} =- 3$
$d_{21} =- 3 + 10 = 7$
$d_{31} =- 3 + 10 - 13 =- 6$
$d_{41} =- 3 + 10 - 13 + 16 = 10$
$d_{n1} =- 3 + \sum_{i=1}^{n-1} (-1)^{i-1}(3i + 7)$

Table 1.3.3.2

The general expression for d_1 becomes:

$$d_{n1} =- 3 + \sum_{i=1}^{n-1} (-1)^{i-1}(3i + 7)$$

Using this, we reconstruct the numerator sequence:

$a_{11} = 6$
$a_{21} = 6 - 3 = 3$
$a_{31} = 6 - 3 + (-3 + 10) = 10$
$a_{41} = 6 - 3 + (-3 + 10) + (-3 + 10 - 13) = 4$
$a_{n1} = 6 + \sum_{i=1}^{n-1} (- 3 + \sum_{j=1}^{i-1} (-1)^{j-1}(3j + 7))$

Table 1.3.3.3

The general form of the numerator sequence is:

$$\{a_n\} = 6 + \sum_{i=1}^{n-1} (- 3 + \sum_{j=1}^{i-1} (-1)^{j-1}(3j + 7))$$

Multiplying this by $\frac{1}{2}$, we obtain the main irregular sequence:

$$\{a_n\} = \frac{1}{2}(6 + \sum_{i=1}^{n-1} (- 3 + \sum_{j=1}^{i-1} (-1)^{j-1}(3j + 7)))$$

Generalized Pattern Equation

In general, such nested summation patterns can be expressed as:

$$\{a_n\} = \frac{1}{c_1}(a \pm \sum_{i=1}^{n-1} (b \pm \sum_{j=1}^{i-1} (-1)^{j-1}(dj \pm c)))$$

From the initial sequence equation, we can derive two distinct forms of irregular nth-term expressions, namely:

$$a_n = a + \sum_{i=1}^{n-1} (-1)^{i-1}(di + b)$$

$$a_n = a + \sum_{i=1}^{n-1} (b + \sum_{j=1}^{i-1} (-1)^{j-1}(dj + c))$$

Additional irregular sequences can be constructed analogously, allowing for the formulation of corresponding nth-term equations.

Building upon these sequences, further irregular patterns may be identified by considering **consecutive gradients**, such as $m_1 = 2$ and $m_2 = 3$. The general form of such a sequence is:

$$\{a_n\} = a + \sum_{i=1}^{n-1} [b + \sum_{j=1}^{i-1} (-1)^j (mj \pm c)]$$

Considering mixed gradients, e.g., $m_1 = 2$ and $m_2 = \frac{1}{3}$ and, where the general equation is:

$$\{a_n\} = \frac{p}{q}(a \pm \sum_{i=1}^{n-1} b \pm \sum_{i=1}^{n-1} \sum_{j=1}^{i-1} (-1)^{j-1} (mj \pm c))$$

By varying the signs of the gradients, one can explore the resulting irregular sequences. For instance, choosing $m_1 = -2$, $m_2 = 3$, or $m_1 = 2$, $m_2 = -\frac{1}{3}$, yields general forms such as:

$$\{a_n\} = a \pm \sum_{i=1}^{n-1} (-1)^{i-1} (d_1 f(i) \pm d_2\, g(i))$$

Or in rational form:

$$\{a_n\} = \frac{p}{q}(a \pm \sum_{i=1}^{n-1} (-1)^i (d_1 f(i) \pm d_2\, g(i))$$

Example Problem and Solution

Problem:
Demonstrate that

$$\frac{1}{3}(9 + \sum_{i=1}^{5} (-1)^i (6f(i) + g(i) + 7)) = 0$$

Given the alignment inequality $g(i) \leq f(i) < n$, and the cardinalities $|n_f| = 2 = |n_g|$.

Solution:
Let the expression inside the summation be denoted as b_i. Then the summation becomes:

$$\frac{1}{3}(9 + \sum_{i=1}^{5} b_i)$$

Where each b_i, is defined as:

$$b_i = (-1)^i(6f(i) + g(i) + 7)$$

Using the alignment and cardinality conditions, we construct the following table (Table 1.3.3.4):

n	1	2	3	4	5
$f(i)$	0	1	1	2	2
$g(i)$	0	0	1	1	2

Table 1.3.3.4

From these values, we compute:

$$b_1 =- 7$$

$$b_2 = 13$$

$$b_3 =- 14$$

$$b_4 = 20$$

$$b_5 =- 21$$

Summing these:

$$\frac{1}{3}(9+[-7+13-14+20-21])=\frac{1}{3}(9-9)=0$$

Hence, the left-hand side equals the right-hand side:

$$LHS = RHS$$

1.4. Alternative Forms of Irregular Sequences

In this section, we examine two distinct sequences that differ significantly from those previously discussed. In the earlier version of these studies, an attempt was made to classify these irregular sequences as either arithmetic or geometric. However, such a classification now appears to be conceptually distant—if not entirely infeasible. Consequently, the prior argument has been omitted and replaced with a revised approach presented herein.

1.4.1. Formulation of another Distinct Sequence

The diagram below illustrates the first five terms of the initial sequence introduced in this section.

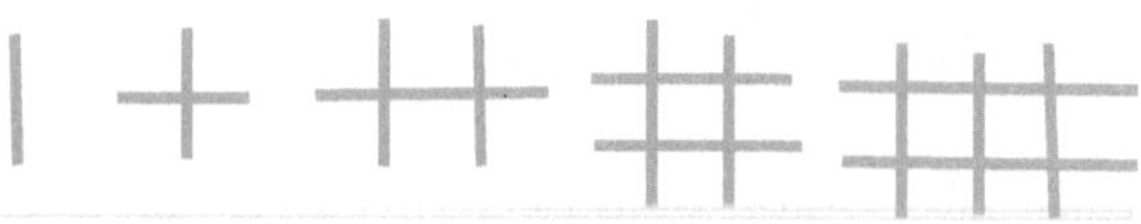

Beginning from the far left, the first diagram contains a single line. As we progress, we count the number of lines preceding any junction in each diagram. For instance, the second diagram contains four lines before a central joint;

the third has seven lines before two joints, and so forth. Enumerating the terms of this sequence yields:

$$\{a_n\} = \{1, 4, 7, 12, 17, 24, 31, 40, ...\}$$

This sequence is clearly irregular, characterized by a non-constant but increasing pattern of common differences. These differences are computed analogously to regular sequences, using the formula $T_{n+1} - T_n$. The resulting sequence of differences is irregular and is:$\{d_n\} = \{3, 3, 5, 5, 7, 7, 9, ...\}$

This can be expressed algebraically as:

$$d_n = 2g(n) + 3 \qquad where\ g(n) \geq 0$$

The following table (Table 1.4.1.1) illustrates the relationship between the control functions $g(n)$ and $f(n)$, the common difference d_n, and the sequence terms a_n (see table below)

n	1	2	3	4	5	6	7
$g(n)$	0	0	1	1	2	2	3
d_n	3	3	5	5	7	7	9
$f(n)$	0	1	1	2	2	3	3
a_n	1	4	7	12	17	24	31

Table 1.4.1.1

To derive a general formula for the sequence, we analyse the interplay between the initial term a, the control functions, and the common difference d_n. Table 1.4.1.2 below outlines how each term is constructed using these components:

Term	Expression	Conditions
T_1	a	$f(1) = 0, g(1) = 0$
T_2	$a + d$	$f(2) = 1, g(2) = 0$
T_3	$2a + d$	$f(3) = 1, g(3) = 1$
T_4	$2a + 2d$	$f(4) = 2, g(4) = 1$
T_5	$3a + 2d$	$f(5) = 2, g(5) = 2$
T_6	$3a + 3d$	$f(6) = 3, g(6) = 2$
T_7	$4a + 3d$	$f(7) = 3, g(7) = 3$

Table 1.4.1.2

It is important to note that the coefficients preceding d correspond to the values of the control function $f(n)$, not $g(n)$. Also, note that in front of a there is a coefficient that means a is multiplied with a control function or expression with it. Based on the observed pattern, the general formula for the nth term of the sequence is:

$$T_n = a + ag(n) + f(n)d_n$$

Generalization and Evaluation of the Irregular Sequence

The expression for the nth term of the sequence is given by:

$$T_n = a + ag(n) + f(n)(2g(n) + 3)$$

This can be algebraically simplified to:

$$T_n = a(g(n) + 1) + f(n)(2g(n) + 3)$$

To validate this formula, consider the case where $n = 7$, with $g(7) = 3$, $f(7) = 3$, and $a = 1$. Substituting these values into the equation:

$$T_7 = (g(7) + 1) + f(7)(2g(7) + 3)$$

$$= (3 + 1) + 3(2(3) + 3) = 31$$

This result aligns with the diagram, which contains 31 lines distributed before and after the joints, confirming the accuracy of the formula.

General Form and Sequence Variability

A broader formulation for generating similar sequences is:

$$T_n = a(g(n) \pm 1) \pm f(n)(2g(n) \pm 3)$$

This general equation allows for the construction of various sequences by adjusting the signs and the cardinality of n. Such flexibility enables the derivation of alternative irregular sequences, as done to some discussed throughout this document.

It is important to note that while common differences play a central role in identifying and constructing irregular sequences, they also appear explicitly within the equations governing these sequences. Moreover, the interaction between control functions—specifically the multiplication of $f(n)$ and $g(n)$—introduces additional complexity and variation.

This particular sequence exhibits behaviour distinct from those examined earlier, especially in its initial terms. Furthermore, the equations defining irregular sequences are not necessarily unique; multiple formulations may describe the same sequence structure.

Additional Observations

An intriguing aspect of this sequence is that the number of joints in each diagram also forms a sequence, with terms:

$$\{0, 1, 2, 4, 6, \ldots\}$$

This emerging pattern bears resemblance to the next sequence to be analysed, suggesting a potential structural or functional relationship worth exploring further.

1.4.2. Analysis and Representation of another Numerical Sequence

Let us now examine a different numerical sequence. We begin by constructing a visual representation of the sequence under consideration.

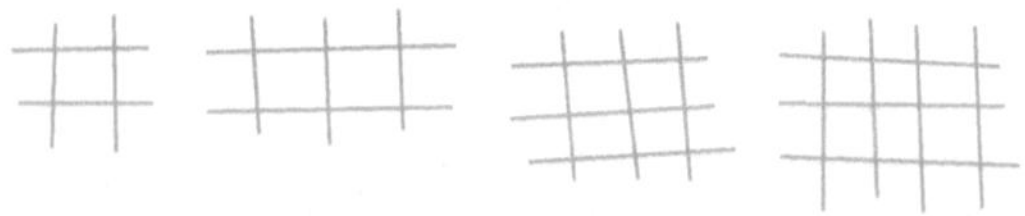

In this diagrammatic approach, we count the number of blocks in each illustration, which collectively define the sequence. By enumerating the sequence through successive block counts from the initial diagram onward, we obtain:

$$\{a_n\} = \{1, 2, 4, 6, 9, 12, 16, 20, 25, 30, \ldots\}$$

Additionally, a common ratio sequence taken from the the terms of the above sequence is given as:

$$\{r_n\} = 2, 2, \frac{3}{2}, \frac{3}{2}, \frac{4}{3}, \frac{4}{3}, \frac{5}{4}, \frac{5}{4}, \frac{6}{5}, \frac{6}{5}, \ldots$$

A recursive formulation for this pattern can be expressed as:

- $T_0 = 1,$
- $T_n = T_{n-1}(\frac{f(n)+1}{f(n)})$
- $, where\ n \geq 1\ and\ |n| = 2.$

The common ratio for the recursive definition is given by:

$$r = \frac{f(n)+1}{f(n)}$$

The values of $f(n)$ for the initial terms are presented in Table 1.4.2.1:

n	1	2	3	4	5	6	7	8
$f(n)$	1	1	2	2	3	3	4	4

Table 1.4.2.1

For instance, on calculating T_9 we use $f(8) = 4$ for $T_8 = 20$ and the common ratio $\frac{5}{4}$, then we compute:

$$T_9 = T_8\left(\frac{5}{4}\right) = 20\left(\frac{5}{4}\right) = 25.$$

We can also express the sequence similarly to the irregular sequence introduced earlier in this section. The reasoning applied previously remains valid here. Table 1.4.2.2 presents the tabulated values for n, control functions and the terms:

n	**1**	**2**	**3**	**4**	**5**	**6**	**7**	**8**	**9**
$f(n)$	1	1	2	2	3	3	4	4	5
$g(n)$	0	1	1	2	2	3	3	4	4
T_n	1	2	4	6	9	12	16	20	25

Table 1.4.2.2: The tabulated values for T_n

From this, we derive the general expression:

$$T_n = f(n)(g(n) + 1), \ where \ |n| = 2$$

The inequality governing the alignment of this function is:

$$g(n) \leq f(n) \leq n$$

For example, when $f(n) = 5$ and $g(n) = 4$, we obtain:

$$T_9 = 5(4 + 1) = 25$$

Thus, any term in the sequence can be determined given the appropriate values of $f(n)$ and $g(n)$. Beyond recursive methods, the expression $T_n = f(n)(g(n) + 1)$ offers an alternative approach. Moreover, it is often possible to derive multiple valid equations for the same sequence or function. This non-uniqueness was also observed in the section on control functions, where a constant function $y = c$ can be equivalently represented by an irregular counterpart $y = f(x)$, provided $f(x)$ is a control function and $|x| = \infty$.

For instance, using the control function formulation $y = 2f(x) + 3$ with $|x| = \infty$ and $f(x) = 1$, we obtain:

$$y = 2(1) + 3 = 5$$

This result mirrors the constant function y = 5 because the substitution is always 1, and this demonstrating the flexibility in representation. In this final example, we considered a sequence with two interrelated sequences:

- A quadratic sequence: $1, 4, 9, 16, 25, \ldots,$ defined by $a_n = n^2$
- An irregular sequence: $2, 6, 12, 20, 30, \ldots$

The second sequence exhibits a pattern of increasing common differences: $4, 6, 8, 10, 12, \ldots,$ which can be described by:

$$d_n = 2n + 2$$

Therefore, the general term of the irregular sequence can be expressed as:

$$\{a_n\} = 2 + \sum_{i=1}^{n-1} (2i + 2)$$

Chapter 2

In this chapter we will introduce two processes that involve the common differences to our sequences that we worked with in chapter 1. They are the addition interchange process and the summation of common differences.

2.1. The Addition Interchange Processes

2.1.1. The First Addition Interchanges Process

The initial interchange process involves two distinct common differences, each associated with a unique control function. Specifically, the presence of two common differences implies the existence of two control functions, with each difference paired exclusively with one function. For instance, the expression:

$$T_n = d_1 f(n) + d_2 g(x) + a$$

illustrates this pairing, d_1 and d_2 are the common differences, and $f(n)$ and $g(n)$ are their respective control functions.

Definition of the Addition Interchange Process

The addition interchange process refers to a systematic method of alternating the addition of two elements, such that each new term is derived by incrementally adding one of the elements to the preceding term. The process begins with an initial term where neither element is added. Subsequently, the first element is added while the second remains unchanged. In the next step, the second element is added, maintaining the first element from the previous term. This alternating pattern continues, with each term incorporating additions in a staggered fashion.

This progression can be illustrated as follows (see Table 2.1.1.1):

a
$a + d_1$
$a + d_1 + d_2$
$a + d_1 + d_1 + d_2$
$a + d_1 + d_1 + d_2 + d_2$
$a + d_1 + d_1 + d_1 + d_2 + d_2$

Table 2.1.1.1: Illustration of the addition interchange process.

To further analyse the additive behaviour, we examine the coefficients of d_1 and d_2 in each term:

Table 2.1.1.2: Numerical representation of the coefficients in the interchange process.

$Numbers\ in\ front\ of\ d_1$	$Numbers\ in\ front\ of\ d_2$
0	0
1	0
1	1
2	1
2	2
3	2

The values in the first column are the control function $f(n)$ values, while those in the second column correspond to $g(n)$. The repetition observed in the $f(n)$ values suggests the necessity of a concatenation operation within the sequence. For example, the sequence of coefficients for d_1 can be expressed as: $0, d_1, d_1, 2d_2, 2d_2, 3d_3, ...$, where the repeated values indicate a structured pattern.

To maintain this repetition, an algebraic concatenation is employed—treating repeated values as elements of a set rather than as strings. For instance, repeating the value 1 twice implies a cardinality of $|n| = 2$, indicating that the value is associated with two successive terms. This approach allows for a formal representation of term positions and their corresponding control values, enhancing the analytical framework of the interchange process.

Algebraic Concatenation and Its Role in the Addition Interchange Process

In order to integrate concatenation within an algebraic framework, it is necessary to redefine the traditional concatenation function—commonly used in computer science—so that its outputs are treated as elements of numerical sets rather than strings. This transformation allows for the application of algebraic operations such as addition and multiplication.

For example, consider the function:

$$f\{n_1, n_2\} \rightarrow a$$

Here, the output set $\{a\} = \{a, a\}$ is valid under the condition that the cardinality of n is 2, i.e., $|n| = 2$. By interpreting the set $\{3, 3\}$ as equivalent to $\{3\}$, we avoid conflating it with the string representation "33". This distinction is crucial for maintaining algebraic integrity, ensuring that repeated numerical values are treated as set elements rather than concatenated strings.

This algebraic interpretation allows us to perform standard mathematical operations on the control function. For instance:

$$f\{n_1, n_2\} \rightarrow a$$
implies that
$$f(n_1) \rightarrow a \text{ and } f(n_2) \rightarrow a$$

thus,

$$f\{n_1, n_2\} \rightarrow a, a$$

In general, we can express this as:

$$f\{n_1, n_2\} \rightarrow X$$

where $\{X = a, a, ...\}$, depending on the cardinality of the input set.

Application of the Addition Interchange Process in Sequences

To demonstrate the practical use of this process, consider the following sequence table:

Term	Expression	Evaluation
T_1	$a + 0 + 0$	$1 + 0 + 0 = 1$
T_2	$a + d_1 + 0$	$1 + 5 + 0 = 6$
T_3	$a + d_1 + d_2$	$1 + 5 + 1 = 7$
T_4	$a + 2d_1 + d_2$	$1 + 2(5) + 1 = 12$
T_5	$a + 2d_1 + 2d_2$	$1 + 2(5) + 2(1) = 13$
T_6	$a + 3d_1 + 2d_2$	$1 + 3(5) + 2(1) = 18$

Table 2.1.1.3: Numerical illustration of the addition interchange process.

From this table, it is evident that a sequence incorporating two distinct common differences can be systematically constructed. The resulting sequence aligns with the formulation introduced in Chapter One, and can be expressed by the equation:

$$T_n = 1 + 5f(n) + g(n)$$

This confirms that the pattern of common differences d_1 and d_2 adheres closely to the principles of the addition interchange process, reinforcing its utility in generating structured numerical sequences.

2.1.2. A Second Addition Interchange Process

The second interchange process is a continuation of the preceding one, characterized by the presence of two distinct common differences too. Consequently, it also necessitates the use of two control functions. This interchange is derived from the sequence:

$$T_n = 1, 3, 5, 7, 8, 10, 12, 14, 15, \ldots$$

The governing equation for this sequence is:

$$T_n = [(n - 1) - f(n)]2 + g(n) + 1$$

From this formulation, it is evident that the first common difference is $d_1 = 2$ and the second common difference is $d_2 = 1$, assuming the initial term is $a = 1$. A more generalized expression for the sequence is:

$$T_n = [(n \pm 1) \pm f(n)]d_1 \pm d_2 g(n) \pm a$$

In this equation:

- The term $((n - 1)d_1)$ accounts for the incremental growth of the sequence.
- The product $(f(n)d_1)$ captures the repetition of values.
- The term $(g(n)d_2)$ introduces additional repetitions.

The following table illustrates the additive interchange pattern, emphasizing the progression of common differences across terms:

Term	***Expression***
T_1	a
T_2	$a + d_1$
T_3	$a + d_1 + d_1$
T_4	$a + d_1 + d_1 + d_1$
T_5	$a + d_1 + d_1 + d_1 + d_2$
T_6	$a + d_1 + d_1 + d_1 + d_1 + d_2$
T_7	$a + d_1 + d_1 + d_1 + d_1 + d_1 + d_2$

Table 2.1.2.1: Term Construction Using Common Differences

This pattern reveals that the repetition of (d_1) occurs every four terms, while (d_2) remains constant during these intervals. The values preceding each common difference—interpreted as control function outputs—are summarized below:

Table 2.1.2.2 – Control Function Values

Number in front of d_1	*Number in front of d_2*
0	0
1	0
2	0
3	0
3	1
4	1
5	1
6	1
6	2

The pattern indicates that while the coefficient of (d_1) increases steadily over four consecutive terms, the coefficient of (d_2) remains unchanged during this interval. Once the coefficient of (d_1) stabilizes (i.e., repeats its last value), the coefficient of (d_2) increments by one. This cyclical behavior continues throughout the sequence.

Key Observations

- For the initial four terms, (d_1) is incremented at each step, starting from $(0d_1)$ and culminating at $(3d_1)$, while (d_2) remains at zero.
- After reaching $(3d_1)$, this value is retained, and (d_2) is introduced.

- The sequence then resumes the incremental pattern for (d_1), progressing to ($6d_1$), which becomes the new base for the next cycle.

This iterative process highlights the role of control functions ($f(n)$) and ($g(n)$) in managing the repetition and growth of terms within irregular sequences. The introduction of these functions facilitates algebraic manipulation and repetition, which would otherwise be challenging to achieve through conventional means. Thus, control functions serve as essential tools for structuring and analysing non-uniform sequences.

What we can learn from these two addition interchanges. We had to introduce the control function ($f(n)$) in order to work with the numbers in front of the d's. introducing the control function enabled us to do the repetition we needed. We did repeat a value or a set of values with the help of this function. We can now see why working with irregular sequences was not so obvious. Mathematically, it was impossible to repeat the values algebraically and experience no problems (but now we can with the help of this control function).

2.2. The Summation of Common Differences

2.2.1. Analysis of a Non-Uniform Sequence via Summation of Common Differences

Consider the sequence defined as follows:

$$\{a_n\} = 9, 2, 15, 1, 21, 0, 27, -1, 33, -2, \dots \qquad \text{(S1)}$$

Upon examining the sequence (S1), it becomes evident that the differences between successive terms are not constant. Instead, they form an alternating sub-sequence. The sequence of these common differences is given by:

$$\{d_n\} = -7, 13, -14, 20, -21, 27, -28, 34, -35, \ldots \quad (S2)$$

To derive a general expression for this sequence of differences (S2), we first consider the absolute values of its terms, resulting in a new sequence:

$$\{+ved_n\} = 7, 13, 14, 20, 21, 27, 28, 34, 35, \ldots$$

Analysing the differences within this positive sequence, we observe two distinct patterns:

- Differences such as $13 - 7 = 6$ and $20 - 14 = 6$
- Differences such as $14 - 13 = 1$ and $21 - 20 = 1$

This indicates the presence of two control functions governing the structure of the sequence. Let the initial term be a = 7, with first and second common differences $d_1 = 6$ and $d_2 = 1$, respectively. The general term of the sequence can then be expressed as:

$$T_n = 6f(n) + g(n) + 7$$

The cardinality of n in the current sequence is 2. The functions $f(n)$ and $g(n)$ are defined as in the table that follows:

n	1	2	3	4	5	6	7	8	9
$f(n)$	0	1	1	2	2	3	3	4	4
$g(n)$	0	0	1	1	2	2	3	3	4
T_n	7	13	14	20	21	27	28	34	35

Table 2.2.1.1: Table showing positive d_n terms

For instance, evaluating T_8 yields:

$$T_8 = 6(4) + 3 + 7 = 24 + 10 = 34$$

To recover the original alternating sequence (S2), we incorporate the alternating sign pattern, resulting in:

$$d_n = (-1)^n(6f(n) + g(n) + 7)$$

We further reconstruct the initial sequence (S1), by performing a cumulative summation of the terms in (S2), beginning with the first term $T_1 = 9$. This process is illustrated in Table 2.2.1.2 below:

$\mathbf{T_1 = 9}$
$\mathbf{T_2 = 9 - 7 = 2}$
$\mathbf{T_3 = 9 - 7 + 13 = 15}$
$\mathbf{T_4 = 9 - 7 + 13 - 14 = 1}$
$\mathbf{T_5 = 9 - 7 + 13 - 14 + 20 = 21}$
$\mathbf{T_6 = 9 - 7 + 13 - 14 + 20 - 21 = 0}$

Table 2.2.1.2: Table showing the cumulative summation of T_n

This summation process can be generalized by the following expression:

$$T_n = 9 + \sum_{i=1}^{n-1} (-1)^i(6f(i) + g(i) + 7)$$

This formula enables the computation of any term in the original sequence (S1) by summing the alternating common differences and adding them to the initial term. In certain contexts, the summation may be omitted, yielding a simplified form:

$$T_n = 9 + (-1)^i(6f(i) + g(i) + 7)$$

More generally, the structure of such sequences can be expressed as:

$$T_n = a \pm (-1)^n(d_1 f(n) \pm d_2 g(n) \pm b)$$

This formulation encapsulates the alternating nature and dual control mechanisms of the sequence.

2.2.2. Nested Summation and Sequence Construction

We now consider a problem involving a hierarchical structure of summations. Let us begin with the sequence defined as:

$$a_n = 6, 3, 10, 4, 14, 5, 18, 6, \ldots$$

The sequence of first-order differences derived from a_n yields an alternating sub-sequence:

$$d_n = -3, 7, -6, 10, -9, 13, -12, \ldots$$

Proceeding further, the sequence of second-order differences—computed from d_n—produces another sub-sequence:

$$d_i = 10, -13, 16, -19, 22, \ldots$$

To formulate an expression for this final sequence, we isolate its positive terms. Although a control function $f(i)$ is not explicitly defined, we may substitute $f(i) = i$, given their equivalence in this context. The positive terms follow the linear relation:

$$|d_i| = 3i + 7$$

Incorporating the alternating sign pattern, we obtain the general expression:

$$d_i = (-1)^{i-1}(3i + 7)$$

To reconstruct the sequence d_n, we sum the terms of d_i and add the result to the initial term of d_n. This yields:

$$d_n =- 3 + \sum_{i=1}^{n-1} (-1)^{i-1}(3i + 7)$$

The following table (next page) illustrates the cumulative construction of d_n:

Term	Value
T_1	-3
T_2	-3 + 10 = 7
T_3	-3 + 10 - 13 = -6
T_4	-3 + 10 - 13 + 16 = 10
T_5	-3 + 10 - 13 + 16 - 19 = -9
T_6	-3 + 10 - 13 + 16 - 19 + 22 = 13

Table 2.2.2.1: Construction of d_n

Having established the expression for the common differences, we now derive the defining equation for the original sequence a_n by summing the values of d_n and adding them to the initial term of a_n:

$$a_n = 6 + \sum_{j=1}^{n-1} [- 3 + \sum_{i=1}^{j-1} (-1)^{i-1}(3i + 7)]$$

The table below presents selected terms of the sequence a_n, computed using the equation for a_n above:

Term	Value
T_1	6
T_2	6 - 3 = 3
T_3	6 - 3 + (-3 + 10) = 10
T_4	6 - 3 + (-3 + 10) + (-3 + 10 - 13) = 4

Table 2.2.2.2: Sequential values of a_n

Chapter 3

In this chapter, we explore irregular functions and functions composed of straight-line segments, particularly those that are interconnected to form piecewise or composite graphs. Building on the foundational concepts introduced in Chapter 1, we now begin to delve deeper into the structural behaviour of linear components within more functions.

3.1. Irregular Functions and Control-Mediated Mapping

A function is formally defined as a relation that assigns to each element x in the domain a unique element y in the codomain. This implies that while multiple values of x may correspond to the same y, no single x may yield more than one distinct output. Thus, the mapping preserves the principle of uniqueness in the output.

In the context of **irregular functions**, this foundational definition remains intact. However, the mapping from (x) to (y) is mediated by one or more **control functions**, which serve as intermediate operators. These control functions do not alter the functional nature of the relation but rather introduce a layered structure in the computation of the dependent variable.

Consider a general form of an irregular function involving two control functions:

$$y = w(x) = d_1 f(x) \pm d_2 g(x) \pm a$$

Here, $f(x)$ and $g(x)$ are control functions, d_1 and d_1 are gradient coefficients, and a is the constant term representing the base value when both control functions vanish. The corresponding flow diagram (Figure 3.1.1) illustrates the sequential dependency: the independent variable x first yields the control values $f(x)$ and $g(x)$, which are then linearly combined via their respective gradients to produce the final output $y = w(x)$.

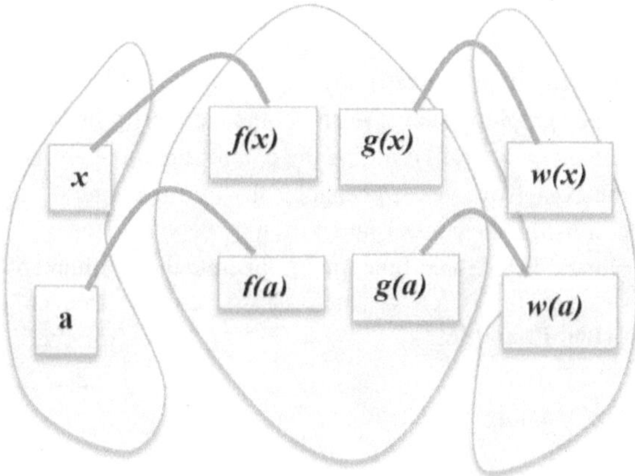

Figure 3.1.1

Role of Control Functions

Control functions occupy a dual role:

- They are **post-independent variables**, derived from x,
- Simultaneously, they are **pre-dependent variables**, feeding into the computation of y.

This intermediary status positions them as both part of the domain and part of the range—analogous to a transitional point on the real line, such as zero, which must be traversed to move between positive and negative values.

Importantly, the presence of control functions does not invalidate the relation as a function. The mapping from (x) to (y) remains well-defined and unambiguous. The control functions merely transform the input prior to final evaluation.

Graphical Representation

When graphing such a relation, one typically plots x against $y = w(x)$, omitting the control functions from the axes. However, if pedagogically useful and non-confusing, the control functions may be visualized alongside the main function to illustrate their influence.

Worked Example

Let us examine the function:

$$y = f(x) + 5g(x) + 3$$

Using the following table (Table 3.1.1):

x	1	2	3	4	5	6	7
$f(x)$	0	1	1	2	2	3	3
$g(x)$	0	0	1	1	2	2	3
y	3	4	9	10	15	16	21

Table 3.1.1: Function values of $y = f(x) + 5g(x) + 3$

This function exhibits **alternating gradient influence**. The gradients are obtained via partial differentiation:

$$m_1 = \frac{\partial y}{\partial f(x)} = 1 \; and \; m_2 = \frac{\partial y}{\partial g(x)} = 5$$

These values indicate that changes in $g(x)$ have a fivefold impact on y compared to changes in $f(x)$. The function increases in a stepwise manner, alternating between the influence of $f(x)$ and $g(x)$, as depicted in Graph 3.1.1.

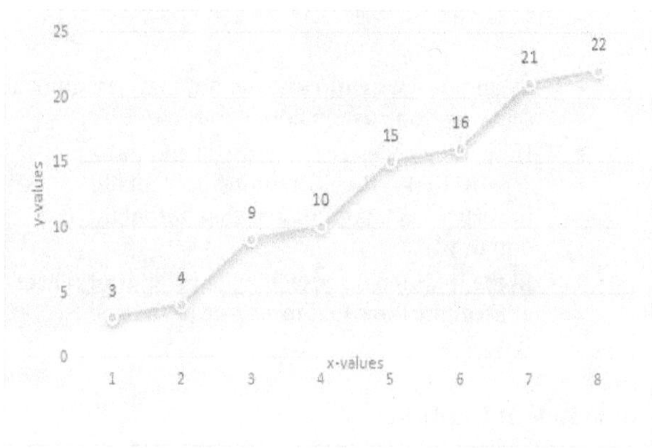

Graph 3.1.1

Function Validity

The graph of $y = w(x)$ satisfies the vertical line test, confirming that each input x maps to a unique output y. Therefore, despite the presence of control functions, the relation qualifies as a function.

3.1.1. The Role of Gradients in Irregular Functions

In our framework of **irregular functions**, gradients play a nuanced and powerful role—they're not just rates of change, but **structural weights** that govern how control functions contribute to the final output. Let's unpack this in layers:

What Are Gradients in This Context?

In the expression:

$$y = w(x) = d_1 f(x) \pm d_2 g(x) \pm a$$

- d_1 and d_2 are **gradients**, but not in the traditional sense of slope over a continuous curve.
- Here, they act as **coefficients** or **partial sensitivities**—they determine how much influence each control function has on the output y.
- In the traditional sense, they are **slopes of pieces of straight lines** that form part of the main function.

Dual Role of Gradients

1. Weighting Mechanism

- Gradients scale the contribution of each control function.
- For example, in$y = f(x) + 5g(x) + 3$, the gradient of 5 means that changes in $g(x)$ affect y five times more than changes in $f(x)$.

2. Directional Control

- Gradients can be **positive, negative, or zero**, allowing for:
    - **Amplification** (positive gradient)
    - **Inversion** (negative gradient)
    - **Suppression** (zero gradient, effectively removing the control function)

Partial Derivatives as Gradient Extractors

We can elegantly use partial derivatives to isolate the gradients:

$$\frac{\partial y}{\partial f(x)} = d_1 \quad \textit{and} \quad \frac{\partial y}{\partial g(x)} = d_2$$

This formalizes the idea that each control function contributes independently to y, and the gradient quantifies that contribution.

Gradient Interchange and Stepwise Behaviour

In our example, the function exhibits **alternating gradients**:

- First, $f(x)$ increases while $g(x)$ remains constant → small step in y
- Then, $g(x)$ increases → large step in y
- This alternation creates a **stepwise growth pattern**, visible in both the table and graph.

This behaviour is **not smooth**, but **discrete and modular**, which is a hallmark of irregular functions.

Boxed Insight: Gradient Functionality in irregular functions

In control-mediated functions of the form $y = w(x) = d_1 f(x) \pm d_2 g(x) \pm a$ gradients $\{d_1, d_1\}$ serve as:

- Weighting coefficients that scale the influence of control functions
- Partial derivatives that quantify sensitivity of y to each control function

- Operators of structural modulation, enabling stepwise, alternating, or suppressed behaviour

Pedagogical Extension

You might consider introducing:

- A **gradient lattice diagram**, showing how combinations of (d_1) and (d_2) affect the shape of (y)
- A **gradient-switching theorem**, formalizing when and how gradient dominance shifts between control functions

3.1.2. Bijectivity of Irregular Functions

To determine whether an irregular function is bijective, we must verify that it satisfies both **injectivity** and **surjectivity**. The approach adopted here draws on the methodology outlined by Mynhardt (2016).

Injectivity

Let w: $A\ to\ B$ be a function. Then w is injective (one-to-one) if and only if, for all $a, a^{'}$ in $Dom(w)$,

$$w(a) = w(a^{'})$$
$$\text{then } a = a^{'}$$

To test injectivity for the irregular function $w(x) = f(x) + 5g(x) + 3$, consider arbitrary elements $x, x^{'} \in Dom(w)$. Suppose:

$$w(x) = w(x^{'})$$

Then

$$f(x) + 5g(x) + 3 = f(x') + 5g(x') + 3$$

Subtracting the constants:

$$f(x) + 5g(x) = f(x') + 5g(x')$$

$$f(x) - f(x') + 5(g(x) - g(x')) = 0$$

This equality holds only if $x = x'$, assuming the control functions f and g are themselves injective or sufficiently distinct. Thus, the function w is injective.

Surjectivity

To establish surjectivity, we must show that for every y in $Codomain(w)$, there exists an x in the Dom (w) such that $w(x) = y$.

Given:

$$y = f(x) + 5g(x) + 3$$

We can rearrange to express $f(x)$ and $g(x)$ in terms of y:

$$f(x) = y - 5g(x) - 3$$

$$\text{and } g(x) = \frac{y - f(x) - 3}{5}$$

Substituting back into the original expression confirms:

$$w(x) = f(x) + 5g(x) + 3 = y$$

Hence, for any y, there exists a corresponding x such that $w(x) = y$, confirming surjectivity.

Conclusion

Since $w(x) = f(x) + 5g(x) + 3$ is both injective and surjective, it is bijective.

Graphing Irregular Functions

When visualizing irregular functions, the graph is typically plotted with x on the horizontal axis and $y = w(x)$ on the vertical axis. The control functions $f(x)$ and $g(x)$ are not plotted directly, as they serve as internal operators rather than independent variables. Thus, the function maintains a one-input, one-output structure, consistent with standard functional behaviour.

Irregular functions, like their regular counterparts, may exhibit many-to-one mappings. In fact, control functions themselves often display many-to-one characteristics, as evidenced in prior examples. Moreover, bijective irregular functions admit well-defined inverses, further reinforcing their functional integrity.

3.1.3. The Summation relation as a Functional Relation

Consider the evaluated summation expression:

$$y = \sum_{i=1}^{4} i^2 = 1 + 4 + 9 + 16 = 30$$

While this yields a single value, it can be generalized into a function by allowing the upper limit to vary:

$$y = \sum_{i=1}^{x} i^2$$

This formulation defines a function from $x \in \mathbb{N}$ to $y \in \mathbb{N}$, where each input x produces a corresponding output y via partial summation. For example:

- $x = 1$: then $y = 1^2 = 1$
- $x = 2$: then $y = 1^2 + 2^2 = 1 + 4 = 5$
- $x = 3$: then $y = 1^2 + 2^2 + 3^2 = 1 + 4 + 9 = 14$

This yields the values in Table 3.1.3.1 below:

x	1	2	3	4	5
y	1	5	14	30	55

Table 3.1.3.1

Graphical Behaviour

The graph of this summation function exhibits a steep, parabolic rise, closely resembling the curve of $y = x^2$, which is its continuous analogue. Compared to the gradient function $y = x^2$, the summation function grows more rapidly due to the cumulative nature of its definition.

This example illustrates that summation relations, when properly structured, can be treated as legitimate functions—particularly when the upper bound is variable and the summation rule is consistent.

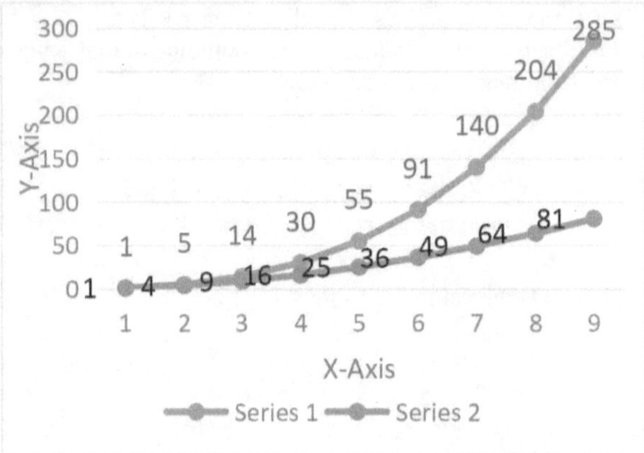

Graph 3.1.3.1: Graph of the summation function compared to $y = x^2$

3.2. Composite Linear Functions and Graphical Structure

Previously, we examined functions such as:

$$y = 5f(x) + g(x) + 1$$

This expression represents a composite function formed by two linear components with gradients of 5 and 1, respectively. These segments are joined to produce a graph that reflects the combined behaviour of both lines. Such constructions exemplify how multiple linear elements can be integrated to form a richer graphical representation.

Further examples of this theory appear in Chapter 4, where we investigate limits and derivatives. There, we encounter functions like:

$$y = 1 + \sum_{i=1}^{x-1} (-1)^i (i - 1)$$

Unlike the previous examples, this function comprises numerous interconnected linear segments. Each term in the summation contributes a distinct linear piece (characterised by a unique gradient), resulting in a graph that is built from many short straight lines. This aggregation of linear components illustrates a broader extension of straight-line theory.

Another illustrative function we considered is:

$$y = 3f(x) + 2$$

Similar to the earlier example $y = 5f(x) + g(x) + 1$, this function is composed of two linear segments with gradients of 0 and 3. These segments are joined to form a graph that reflects the piecewise nature of the function.

Throughout this book, we encounter numerous functions that fall within what we term the "extended theory of straight lines." These functions not only utilize linear components but also demonstrate how such elements can be systematically combined to model more intricate behaviours. In doing so, they enrich our understanding of linearity and its applications in mathematical modelling.

3.3. The Other Part of the Theory of Straight Lines

This part has been taken from the previous preprint article *The Extension of the Theory of Straight Lines and Second Degree Irregular Functions* by (Mbele, 2025). The sequences (when taken as functions) examined thus far fall within the framework of linear theory. This classification arises from the observation that, when plotted, their graphical representations consist of multiple straight-line segments joined to form a composite figure. We now turn our attention to the following sequence (which will be treated as a function, without losing sight of the sequential values):

$$\{a_n\} = 1, 2, 3, 3, 2, 1, 1, 2, 3, 3, 2, 1, 1, ...$$

This sequence exhibits a recurring pattern of integers ranging from 1 to 3, alternating between ascending and descending sub-sequences. Specifically, the subsequence 1, 2, 3 is strictly increasing, while 3, 2, 1 is strictly decreasing. These alternating patterns repeat indefinitely, and our objective is to determine a general expression for the nth term without resorting to full enumeration.

To analyse the structure, we consider the sequence of first-order differences:

$$\{d_n\} = 1, 1, 0, -1, -1, 0, 1, 1, 0, -1, -1, \ldots$$

This sequence reveals three distinct difference values (gradients to the corresponding function): $+1$, 0, and -1. Consequently, we introduce three control functions to model the behaviour:

- $f(n)$ corresponds to $d = +1$
- $g(n)$ corresponds to $d = 0$
- $h(n)$ corresponds to $d = -1$

The above control functions are to be treated as functions especially when working with them individually. Given that the initial term is $a = 1$, and that $g(n)$ contributes no net change, the general term can be expressed as:

$$T_n = a + f(n) - h(n) \qquad eq. 3.3.1$$

This formulation is illustrated in Table 3.3.1:

n	a	$f(n)$	$g(n)$	$-h(n)$	T_n
1	1	0	0	0	1
2	1	1	0	0	2
3	1	2	0	0	3
4	1	3	0	1	3
5	1	4	0	3	2
6	1	5	0	5	1
7	1	6	0	6	1
8	1	7	0	6	2
9	1	8	0	6	3
10	1	9	0	7	3
11	1	10	0	9	2
12	1	11	0	11	1

Table 3.3.1: Construction of T_n via control functions

It is important to note that each term T_n includes the initial value $a = 1$ and the contribution from $f(n)$. The control function $h(n)$ operates over a domain with cardinality $|n_h| = 3$, while $f(n)$ spans a domain with cardinality $|n_f| = 1$.

A key question arises: how are the values of $h(n)$ determined, given the irregularity of both the primary sequence and its associated sub-sequences? To address this, we examine the values of $h(n)$ prior to applying the gradient $d_3 = -1$ i.e. we consider absolute values of $h(n)$:

$$h(n) = 0, 0, 0, 1, 3, 5, 6, 6, 6, 7, 9, 11, \ldots$$

Computing the first-order differences of $h(n)$ yields the following sequence:

$$\{d_s\} = 0,0,1,2,2,1,0,0,1,2,2,1,...$$

This derived difference sequence provides insight into the internal structure of $h(n)$, which will be instrumental in constructing a closed-form expression or recursive formulation for T_n. To model the irregular sequence of common differences, we have defined it as a sequence d_s, representing the **sequential common difference**. This sequence exhibits a repeating pattern:

$$d_s = \{0, 0, 1, 2, 2, 1\}$$

which recurs every six terms. This sequence is also constant, increasing and decreasing and it repeats the same sequence of values, viz. 0, 0, 1, 2, 2, 1, ... up to infinity.

To systematically generate d_s, we introduce two control functions—denoted by capital letters to distinguish them from previously defined control functions $f(n)$, $g(n)$, $h(n)$. These new control functions, $F(n)$ and $G(n)$, are designed to alternate between the values 0 and 1 with a cardinality of 3, meaning each value appears three times before the pattern resets.

Table 3.3.2: Generating d_s from Control Functions

d_s	$F(n)$	$G(n)$
1	1	0
0	0	0
0	0	0
1	0	1
2	1	1
2	1	1
1	1	0

From this table, we observe that:

$$d_s(n) = F(n) + G(n)$$

This formulation allows us to capture the repeating structure of d_s using simple binary control functions. But what are the equations to calculate both $F(n)$ and $G(n)$? Treating them as functions, enables us to call them $F(x)$ and $G(x)$ by replacing the variable n with x.

Analytical Derivation of Control Functions F(x) and G(x)

The functions $F(x)$ and $G(x)$ exhibit distinct characteristics, primarily due to their binary output values—restricted to 0 and 1. Consequently, it is imperative to formulate explicit equations for these functions to facilitate the determination of related variables such as y in functions or T_n in sequences. It is established that the cardinality of x for both F and G is 3.

To initiate this analytical endeavour, we examine a tabulated representation of $F(x)$, designated as Table 3.3.4:

x	1	2	3	4	5	6	7	8	9
$F(x)$	1	0	0	0	1	1	1	0	0

Table 3.3.3

From this dataset, we aim to derive three distinct equations correlating x with $F(x)$, as suggested by the cardinality. The general form of the equation is:

$$x = 3F(x) + \sigma_i(\phi), \quad i = 1, 2, 3$$

To determine $\sigma_i(\phi)$, we extract a subset of values from Table 3.3.4, forming Table 3.3.5:

x	1	4	7	10	13	16
$F(x)$	1	0	1	0	1	0

Table 3.3.4

Using the equation $x = 3F(x) + \sigma_1(\phi)$, we isolate $\sigma_1(\phi)$ as:

$$\sigma_1(\phi) = x - 3F(x)$$

This yields the sequence: $-2, 4, 4, 10, 10, 16, ...$ These values correspond to an irregular function $\sigma_1(\phi)$, summarized in Table 3.3.6:

ϕ	1	2	3	4	5	6
$v(\phi$	1	2	2	3	3	4
$\sigma_1(\phi)$	-2	4	4	10	10	16

Table 3.3.5

From prior foundational concepts, the functional form of $\sigma_1(\phi)$ is:

$$\sigma_1(\phi) = 6v(\phi) - 8$$

The control function $v(\phi)$ relates to ϕ via:

- $\phi = 2v(\phi) - 1$ for odd ϕ
- $\phi = 2v(\phi) - 2$ for even ϕ

Additionally, the relationship between x and ϕ is:

$$x = 3\phi - 2$$

Applying similar methodology to $\sigma_2(\phi)$, we derive:

$$\sigma_1(\phi) = 6v(\phi) - 4$$

With the associated relationships:

- $\phi = 2v(\phi) - 1$
- $\phi = 2v(\phi)$
- $x = 3\phi - 1$

For $\sigma_3(\phi)$, the equation is:

$$\sigma_3(\phi) = 6v(\phi) - 3$$

The associated relationships are

- $\phi = 2v(\phi) - 1$
- $\phi = 2v(\phi)$
- $x = 3\phi$

The comprehensive expression for x is thus:

$$x = 3F(x) + \begin{cases} \sigma_1(\phi) = 6v(\phi) - 8 \; if \begin{cases} \phi = 2v(\phi) - 1 \\ \phi = 2v(\phi) - 2 \end{cases} when \; x = 3\phi - 2 \\ \sigma_2(\phi) = 6v(\phi) - 4 \; if \begin{cases} \phi = 2v(\phi) - 1 \\ \phi = 2v(\phi) \end{cases} when \; x = 3\phi - 1 \\ \sigma_3(\phi) = 6v(\phi) - 3 \; if \begin{cases} \phi = 2v(\phi) - 1 \\ \phi = 2v(\phi) \end{cases} when \; x = 3\phi \end{cases}$$

Example 3.3.1: Determining $F(11)$

To evaluate $F(11)$, we test each case sequentially:

1. $11 = 3\phi - 2 \Rightarrow \phi = \frac{13}{3} \rightarrow$ Not valid
2. $11 = 3\phi - 1 \Rightarrow \phi = 4 \rightarrow$ Valid

Now, determine $v(4)$:

- $4 = 2v(4) - 1 \Rightarrow v(4) = \frac{5}{2} \rightarrow$ Invalid
- $4 = 2v(4) \Rightarrow v(4) = 2 \rightarrow$ Valid

Then:

$$\sigma_2(4) = 6v(4) - 4 = 6(2) - 4 = 8$$

From Table 3.3.7:

ϕ	**1**	**2**	**3**	**4**	**5**	**6**
$v(\phi)$	1	1	2	2	3	3
$\sigma_2(\phi)$	2	2	8	8	14	14

Table 3.3.6

Finally:

$$x = 3F(x) + \sigma_2(4)$$

$$11 = 3F(11) + 8 \Rightarrow F(11) = \frac{3}{3} = 1$$

This confirms the validity of the derived equation and its applicability to compute $F(x)$ for arbitrary values of x.

Derivation of the Control Functions $G(x)$

Given the structural similarity between the derivation processes for $G(x)$ and $F(x)$, we proceed by directly stating the general form of the equation for $G(x)$:

$$x = 3G(x) + \begin{cases} \sigma_1(\phi) = 6v(\phi) - 5 \; if \begin{cases} \phi = 2v(\phi) - 1 \\ \phi = 2v(\phi) \end{cases} when \; x = 3\phi - 2 \\ \sigma_2(\phi) = 6v(\phi) - 4 \; if \begin{cases} \phi = 2v(\phi) - 1 \\ \phi = 2v(\phi) \end{cases} when \; x = 3\phi - 1 \\ \sigma_3(\phi) = 6v(\phi) - 3 \; if \begin{cases} \phi = 2v(\phi) - 1 \\ \phi = 2v(\phi) \end{cases} when \; x = 3\phi \end{cases}$$

Example 3.3.2: Evaluation of $G(11)$

To determine $G(11)$, we begin by solving for ϕ using the outermost expressions:

- $11 = 3\phi - 2 \Rightarrow \phi = \frac{13}{3} \rightarrow$ Not valid
- $11 = 3\phi - 1 \Rightarrow \phi = 4 \rightarrow$ Valid

Next, we test whether $\phi = 4$ satisfies the control function conditions:

- $4 = 2v(4) - 1 \Rightarrow v(4) = \frac{5}{2} \rightarrow$ Invalid
- $4 = 2v(4) \Rightarrow v(4) = 2 \rightarrow$ Valid

Substituting into the equation for $\sigma_2(\phi)$:

$$\sigma_2(4) = 6v(4) - 4 = 6(2) - 4 = 8$$

Now, solving for $G(11)$:

$$11 = 3G(11) + 8 \Rightarrow G(11) = \frac{3}{3} = 1$$

The result confirms the validity of the equation.

To compute the cumulative effect of d_s on the sequence T_n, we introduce an auxiliary accumulator function $H(n)$, which updates based on the previous value of $h(n)$ and the current $d_s(n)$. Specifically:

$$h(n) = h(n-1) + d_s(n) = H(n) + d_s(n)$$

This recursive relationship enables us to track the evolving structure of the target function y or sequence T_n.

Table 3.3.7: Recursive Construction of $h(n)$

T_n	$H(n)$	$d_s(n)$	$h(n)$
T_1	−1	1	0
T_2	0	0	0
T_3	0	0	0
T_4	0	1	1
T_5	1	2	3
T_6	3	2	5
T_7	5	1	6

Note that some control functions, such as $g(n)$ from Table 3.3.1, may vanish in the final formulation due to their constant zero values. This simplification is part of the broader strategy to isolate the essential components of the main equation. From the preceding tables and analysis, we observe that the control function $h(n)$ can be expressed as:

$$h(n) = H(n) + d_s(n) = H(n) + (F(n) + G(n))$$

Recalling the earlier formulation ($eq.$ 3.3.1):

$$T_n = a + f(n) - h(n)$$

we substitute the expression for h(n) to obtain:

$$T_n = a + f(n) - \left(H(n) + \left(F(n) + G(n)\right)\right) \qquad eq. 3.3.2$$

This equation captures the structure of the main sequence:

$$\{a_n\} = 1, 2, 3, 3, 2, 1, 1, 2, 3, 3, 2, 1, 1, \ldots$$

Analysis of the Function $H(x)$

Unlike $F(x)$ and $G(x)$, the function $H(x)$ exhibits both increasing and uniform behavior, suggesting a more complex structure. Table 3.3.8 presents the initial values:

x	1	2	3	4	5	6	7	8	9
$H(x)$	-1	0	0	0	1	3	5	6	6

Table 3.3.8

To derive the governing equations, we analyse three sub-sequences.

Table 3.3.9: First Sub-sequence: Irregular Pattern

x	1	4	7	10	13	16
$H(x)$	-1	0	5	6	11	12

This sequence exhibits alternating common differences of 1 and 5. Introducing control variables $u(\phi)$ and $v(\phi)$, we construct Table 3.3.10:

ϕ	1	2	3	4	5	6	7
$u(\phi)$	0	1	1	2	2	3	3
$v(\phi)$	0	0	1	1	2	2	3
$\sigma_1(\phi)$	-1	0	5	6	11	12	17

Table 3.3.10

From this, we derive:

$$\sigma_1(\phi) = u(\phi) + 5v(\phi) - 1$$

Second Sub-sequence: Similar Irregularity

x	2	5	8	11	14	17
$H(x)$	0	1	6	7	12	13

Table 3.3.11

Table 3.3.12 provides the control structure:

ϕ	1	2	3	4	5	6
$u(\phi)$	0	1	1	2	2	3
$v(\phi)$	0	0	1	1	2	2
$\sigma_2(\phi)$	0	1	6	7	12	13

Table 3.3.12

The corresponding equation is:

$$\sigma_2(\phi) = u(\phi) + 5v(\phi)$$

Third Sub-sequence: Regular Pattern

x	3	6	9	12	15	18
$H(x)$	0	3	6	9	12	15

Table 3.3.13

This sequence is linear and yields:

$$\sigma_3(\phi) = 3\phi - 3$$

The relationship between n and ϕ for the different $\sigma(\phi)$ are as follows:

- $n = 3\phi - 2$ for the first equation
- $n = 3\phi - 1$ for the second equation
- $n = 3\phi$ for the last one

Final Composite Equation for $H(x)$

$$H(x) = \begin{cases} \sigma_1(\phi) = u(\phi) + 5v(\phi) - 1 \; if \; x = 3\phi - 2 \\ \sigma_2(\phi) = u(\phi) + 5v(\phi) \; if \; x = 3\phi - 1 \\ \sigma_3(\phi) = 3\phi - 3 \; if \; x = 3\phi \end{cases}$$

Example 3.3.3: Evaluation of $H(11)$

Testing the relationships:

- $11 = 3\phi - 2 \Rightarrow \phi = \frac{13}{3} \rightarrow$ Invalid
- $11 = 3\phi - 1 \Rightarrow \phi = 4 \rightarrow$ Valid

From Table 3.3.12:

- $u(4) = 2, v(4) = 1$

$$\sigma_2(4) = u(4) + 5v(4) = 2 + 5(1) = 7$$

Thus:

$$H(11) = 7$$

This confirms the accuracy and applicability of the derived equations for computing $H(x)$.

Dependency on tables

Let us verify Equation (3.3.2) by computing specific terms using values from Tables 3.3.1 and 3.3.7.

$T_n = a + f(n) - \left(H(n) + \left(F(n) + G(n)\right)\right)$ is the equation (3.3.2) for quick reference.

Example 3.3.4: Computing T_7

Given:

- $a = 1$
- $f(7) = 6$
- H(7)=5
- $F(7) = 1$
- $G(7) = 0$

Substituting into Equation (3.2.2):

$$T_7 = 1 + 6 - (5 + 1 + 0) = 1$$

This matches the 7th term in the sequence, confirming the validity of the formulation.

Example 3.3.5: Computing T_{10}

Given:

- $a = 1$
- $f(10) = 9$
- $H(10) = 6$
- $F(10) = 0$
- $G(10) = 1$

Substituting:

$$T_{10} = 1 + 9 - (6 + 0 + 1) = 3$$

Again, this result aligns with the 10th term of the sequence.

It is noteworthy that the dependency on tables can make the job of finding T_n really difficult. So we need to work with equations that can help us find values a bit simpler or easily.

We now simplify the expression by substituting $a = 1$ and $f(n) = n - 1$, yielding:

$$T_n = 1 + (n - 1) - \big(H(n) + (F(n) + G(n))\big)$$

$$T_n = n - \big(H(n) + (F(n) + G(n))\big) \qquad eq.\ 3.3.3$$

To verify Equation (3.3.3), consider n = 10 (again):

- $H(10) = 6$
- $F(10) = 0$
- $g(10) = 1$

Then:

$$T_{10} = 10 - (6 + 0 + 1) = 3$$

This confirms the correctness of the simplified formulation. The functions $F(n)$ and $G(n)$ are specialised control functions that alternate between the values 0 and 1. Unlike traditional control functions, these operate over blocks of three terms, reflecting a cardinality of 3.

To assist with computation and visualization, we present the following table:

n	**1**	**2**	**3**	**4**	**5**	**6**	**7**	**8**	**9**
a	1	1	1	1	1	1	1	1	1
$f(n)$	0	1	2	3	4	5	6	7	8
$F(n)$	1	0	0	0	1	1	1	0	0
$G(n)$	0	0	0	1	1	1	0	0	0
$H(n)$	-1	0	0	0	1	3	5	6	6
$h(n)$	0	0	0	1	3	5	6	6	6
T_n or a	1	2	3	3	2	1	1	2	3

Table 3.3.14: Graph showing the control function values and resulting sequence terms.

Establishing Functional Relationships for Control Functions $u(\phi)$ and $v(\phi)$

As the value of x increases, determining the control functions $u(\phi)$ and $v(\phi)$ via tabular lookup becomes increasingly impractical. Therefore, it is essential to derive explicit relationships between ϕ and these control functions to facilitate computation for large x.

Assuming familiarity with deriving equations from tabular data, we present the relevant tables and the equations inferred from them.

Relationship Between ϕ and $u(\phi)$

Examine on the next table:

Table 3.3.15:

ϕ	1	2	3	4	5	6	7	8
$u(\phi)$	0	1	1	2	2	3	3	4

We extract two subsets:

- **Table 3.3.16** (Odd ϕ):

ϕ	1	3	5	7	9	11
$u(\phi)$	0	1	2	3	4	5

→ Equation: $\phi = 2u(\phi) + 1$

- **Table 3.3.17** (Even ϕ):

ϕ	2	4	6	8	10	12
$u(\phi)$	1	2	3	4	5	6

→ Equation: $\phi = 2u(\phi)$

Relationship Between ϕ and $v(\phi)$

- From a table analogous to Table 3.3.16:

→ Equation: $\phi = 2v(\phi) + 1$

Table 3.3.18: (Even ϕ)

• ϕ	2	4	6	8	10	12	14
$v(\phi)$	0	1	2	3	4	5	6

→ Equation: $\phi = 2v(\phi) + 2$

These relationships apply to both $\sigma_1(\phi)$ and $\sigma_2(\phi)$, while $\sigma_3(\phi)$ remains unaffected due to its regularity.

Revised Equation for $H(x)$

$$H(x) = \begin{cases} \begin{matrix} \sigma_1(\phi) = u(\phi) + 5v(\phi) - 1, \\ \sigma_2(\phi) = u(\phi) + 5v(\phi), \end{matrix} \begin{cases} \phi = 2u(\phi) + 1, \\ \phi = 2u(\phi), \\ \phi = 2v(\phi) + 1, \\ \phi = 2v(\phi) + 2, \end{cases} \begin{matrix} x = 3\phi - 2 \\ x = 3\phi - 1 \end{matrix} \\ \sigma_3(\phi) = 3\phi - 3, \quad x = 3\phi \end{cases}$$

Example 3.3.6: Computing $H(11)$ Without Tables

From prior analysis (see example 3.3.3), $x = 11$ satisfies $x = 3\phi - 1$, yielding $\phi = 4$. We now determine $u(4)$ and $v(4)$:

- Test $\phi = 2u(\phi) + 1$: $4 = 2u(\phi) + 1 \Rightarrow u(\phi) = \frac{3}{2}$ → Invalid
- Test $\phi = 2u(\phi)$: $u(\phi) = 2$ → Valid
- Test $\phi = 2v(\phi) + 1$: $v(\phi) = \frac{3}{2}$ → Invalid
- Test $\phi = 2v(\phi) + 2$: $v(\phi) = 1$ → Valid

Thus:

$$H(11) = u(4) + 5v(4) = 2 + 5(1) = 7$$

This result aligns with previous findings, confirming the validity of the derived equations.

Example 3.3.7: Computing y for $x = 95$

We seek $y(95)$ using:

$y = x - (H(x) + F(x) + G(x))$ or $y = f(x) - h(x) + 1$

Step 1: Determine $\boldsymbol{\phi}$

- $95 = 3\phi - 2 \Rightarrow \phi = \frac{97}{3} \rightarrow$ Invalid
- $95 = 3\phi - 1 \Rightarrow \phi = 32 \rightarrow$ Valid

Step 2: Compute $H(95)$

- $\phi = 2u(\phi) + 1 \Rightarrow u(\phi) = \frac{31}{2} \rightarrow$ Invalid
- $\phi = 2u(\phi \Rightarrow u(\phi) = 16 \rightarrow$ Valid
- $\phi = 2v(\phi) + 1 \Rightarrow v(\phi) = \frac{31}{2} \rightarrow$ Invalid
- $\phi = 2v(\phi) + 2 \Rightarrow v(\phi) = 15 \rightarrow$ Valid

$$H(95) = u(32) + 5v(32) = 16 + 75 = 91$$

Step 3: Compute $F(95)$

Using the same $\phi = 32$:

- $\phi = 2v(\phi) - 1 \Rightarrow v(\phi) = \frac{33}{2} \rightarrow$ Invalid
- $\phi = 2v(\phi) \Rightarrow v(\phi) = 16 \rightarrow$ Valid

$$\sigma_2(\phi) = \sigma_2(32) = 6v(\phi) - 4 = 96 - 4 = 92$$

$$\therefore 95 = 3F(95) + \sigma_2(32) \Rightarrow 95 = 3F(95) + 92 \Rightarrow F(95) = 1$$

Step 4: Compute $G(95)$

Since $G(95)$ shares the same structure as $F(95)$, we conclude:

$$G(95) = 1$$

Final Computation:

$$y = x - (H(x) + F(x) + G(x)) = 95 - (91 + 1 + 1) = 95 - 93 = 2$$

Or

$$y = f(x) - h(x) + 1 = 94 - 93 + 1 = 2$$

Both approaches yield consistent results, validating the methodology for computing $y(x)$ for large values of x.

We now summarize the four key sequences derived from our analysis, each exhibiting distinct structural properties: Starting with sequence a_n which is the main sequence,

$$\{a_n\} = 1, 2, 3, 3, 2, 1, 1, 2, 3, 3, \ldots$$

- **Behaviour**: Increasing → Constant → Decreasing

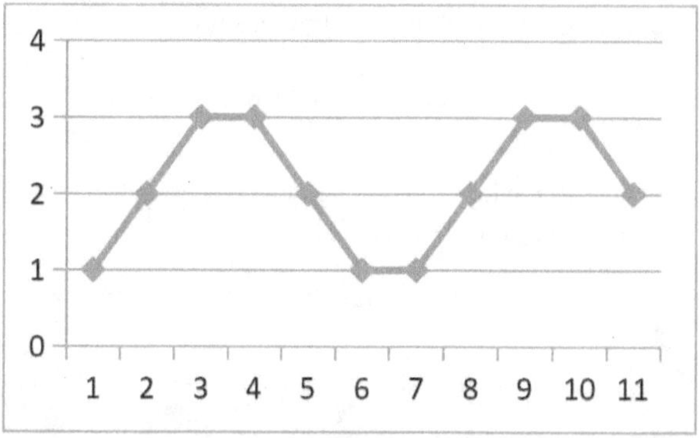

Graph 3.3.1: Part graph of the main sequence

Sequence a_{n2}: Control Function $h(n)$

$$\{a_{n2}\} = 0, 0, 0, 1, 3, 5, 6, 6, 6, 7, 9, 11, 12, \ldots$$

- **Behaviour**: Constant → Increasing

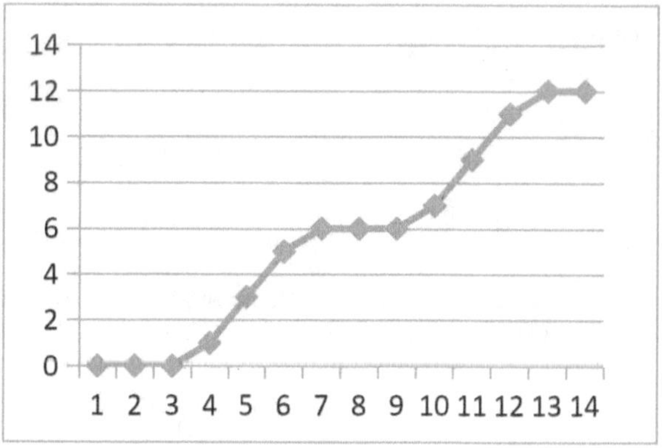

Graph 3.3.2: Sequence for $h(n)$

Sequencea_{n3} : Alternating Control Function

$$\{a_{n3}\} = 0, 0, 0, 1, 1, 1, 0, 0, 0, 1, 1, 1, 0, ...$$

- **Behavior**: Alternating blocks of 0s and 1s
- **Common Differences**: 0, 1, -1

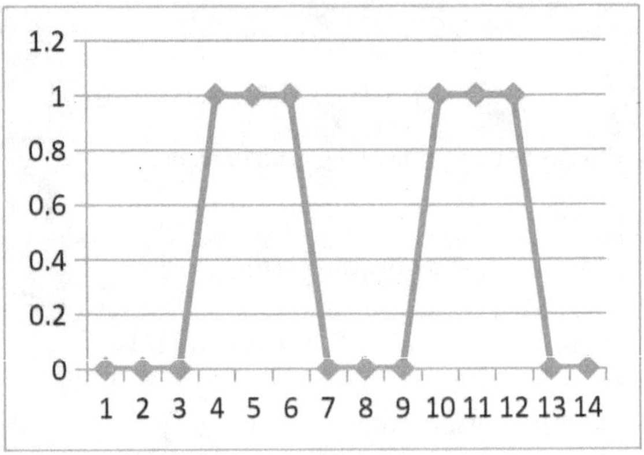

Graph 3.3.3: Switch function representation

Sequence a_{n4}: Repeating Pattern

$$\{a_4\} = 0,0, 1, 2, 2, 1, 0, 0, 1, 2, 2, 1, ...$$

- **Behavior**: Constant → Increasing → Decreasing (repeating)

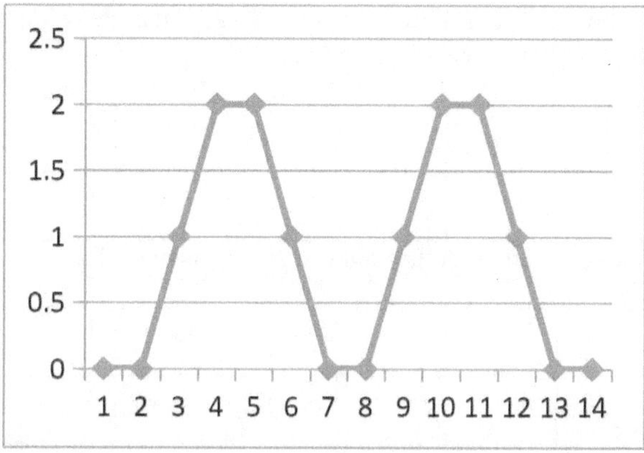

Graph 3.3.4: Graph for the repeating pattern

Finding the biggest nth term (Alternatively)

To simplify matters a little bit, let us workout the nth term of the sequence $a_n = 1, 2, 3, 3, 2, 1, 1, 2, \ldots$ a little bit differently. This time we will try to find the biggest nth term of the sequence.

x	1	2	3	4	5	6	7	8	9
y	1	2	3	3	2	1	1	2	3

Table 3.3.19: Regular table for the function y

Can we find a regular way to define this current function? The function has the values of sequence a_n as its output values. Since the cardinality of x is equal to 3, then we will have the gradient of the defining sub-functions equal to 3. This will enable us to get from the sequence a_n, the following three sub-functions output values that are unique (we enumerate them in table form).

x	1	4	7	10	13
y_1	1	3	1	3	1

Table 3.3.20: First function with binary output values

x	2	5	8	11	14
y_2	2	2	2	2	2

Table 3.3.21: Second function with a single output value repeating

x	3	6	9	12	15
y_3	3	1	3	1	3

Table 3.3.22: Third function with binary output values too

From Table 3.2.7, we have a sub-function that is defined by the equation:

$$x = 3y_1 + \begin{cases} 6\beta - 11 & if\ y_1 = 1 \\ 6\beta - 8 & if\ y_1 = 3 \end{cases} \qquad when\ \beta \geq 1$$

We will as well get from Table 3.2.8, the equation defining the sub-function as:

$$x = 3y_2 + (3\beta - 7) \qquad if\ \beta \geq 1$$

Lastly, we will find the equation for the sub-function values in Table 3.2.9 as:

$$x = 3y_3 + \begin{cases} 6\beta - 3 & if\ y_3 = 3 \\ 6\beta & if\ y_3 = 1 \end{cases}$$

You will notice that the output for equation of Table 3.2.7 has binary output values which interchange upon substitution in the equation. This is the same for the output values for equation of Table 3.2.9.

Example 3: Find the value of y if the value of $x = 93$?

We will use the above three equation to answer our question. First we need to know which equation to use to find y (this is because y can be any of the three values y_1, y_2 and y_3). Note in advance that we will not accept fraction as we work with natural numbers and not with rational numbers. We are going to check the x values in order to see where 93 falls. In Table 3.2.7, the values of x are regularly defined by the equation:

$$a_{n1} = 3n - 2$$

Thus if we substitute $a_{n1} = 93$, we get

$$93 = 3n - 2$$

$$95 = 3n$$

$$n = \frac{95}{3}$$

In Table 3.2.8, the values of x are regularly defined by the equation:

$$a_{n2} = 3n - 1$$

This when using the same value of $a_{n2} = 93$, we obtain

$$93 = 3n - 1$$

$$94 = 3n$$

$$n = \frac{94}{3}$$

Lastly, we use the equation $a_{n3} = 3n$ and substitute $a_{n3} = 93$ to get

$$93 = 3n$$

$$n = \frac{93}{3} = 31$$

As said before, we will only accept the last value (it satisfy the equation) which means we are going to use the third equation, viz.

$$x = 3y_3 + \begin{cases} 6\beta - 3 & if\ y_3 = 3 \\ 6\beta & if\ y_3 = 1 \end{cases}$$

We start with the first part of the equation which says

$$x = 3y_3 + 6\beta - 3 \qquad if\ y_3 = 3$$

In this equation we substitute $x = 93$ and $y_3 = 3$ to obtain

$$93 = 3(3) + 6\beta - 3$$

$$96 - 9 = 6\beta$$

$$\beta = \frac{87}{6} = 14.5$$

We will not accept this value because it does not satisfy the equation. We lastly check the second equation which says

$$x = 3y_3 + 6\beta \qquad if\ y_3 = 1$$

Substituting $x = 93$ and $y_3 = 1$, we get

$$93 = 3(1) + 6\beta$$

$$6\beta = 90$$

$$\beta = \frac{90}{6} = 15$$

Therefore, the last equation is satisfied and the $y = y_3 = 1$ at $x = 93$.

Chapter 4

4.1. The Derivatives

4.1.1. Partial Derivatives of Irregular Functions

- The process of differentiating irregular functions closely parallels that of regular functions, with the derivative still representing a gradient. However, when dealing with summation-based irregular functions, our approach diverges from traditional calculus methods—though not without some notable similarities. This divergence arises primarily because the functions under consideration are composed of linear segments, causing them to behave like straight lines. Consequently, the techniques employed here align with both algebraic and graphical interpretations of linear behaviour.

Importantly, the irregular functions examined in this text do possess derivatives, contrary to common expectations. Nonetheless, it remains true that many other irregular functions may lack differentiability.

Consider an irregular function defined by a single control function, such as $y =- 3f(x) + 2$. Its derivative is straightforwardly -3, mirroring the gradient of its regular counterpart:

$$\frac{dy}{dx} = \frac{dy}{df(x)} =- 3$$

In this framework, differentiation is performed with respect to $f(x)$, analogous to differentiating with respect to x in regular calculus. Remember that this is the power rule method of differentiating. For example:

$$\frac{dy}{dx} = (1 \times -3)x^{1-1} + 0 =-3$$

$$\frac{dy}{df(x)} = (1 \times -3)f^{1-1}(x) + 0 =-3$$

For further reference on regular function derivatives, consult standard calculus texts such as Stewart (2008), Anton (2012), or Thomas (2014).

Using the gradient method yields the same result as the power rule method ($P_1(1, -1)$ and $P_2(0, 2)$):

$$m = \frac{y_2 - y_1}{x_2 - x_1} = \frac{2 - (-1)}{0 - 1} =-3$$

While the calculation of gradients is similar for both regular and irregular functions, the latter requires that the two points used be consecutive. In the current example, the domain of x has cardinality 2, resulting in one gradient of −3 and another of 0. These alternating gradients produce a stepwise graph, with each gradient appearing in succession as x progresses.

To reflect this dual-gradient structure, we express the corresponding irregular function as:

$$y =-3f(x) + 0g(x) + 2$$

This formulation ensures that the zero-gradient component is not overlooked, even though it is not explicitly visible in the simpler expression $y =-3f(x) + 2$. It also underscores the importance of selecting x-values carefully, particularly in relation to cardinality and repetition.

The zero-gradient is evident in the graph of this function (see Graph 4.1.1), and its derivative with respect to $g(x)$ is:

$$\frac{dy}{dg(x)} = 0 + 0 = 0$$

Here, $f(x)$ has been treated as a constant, and the derivative represents the "invisible" component of the original function.

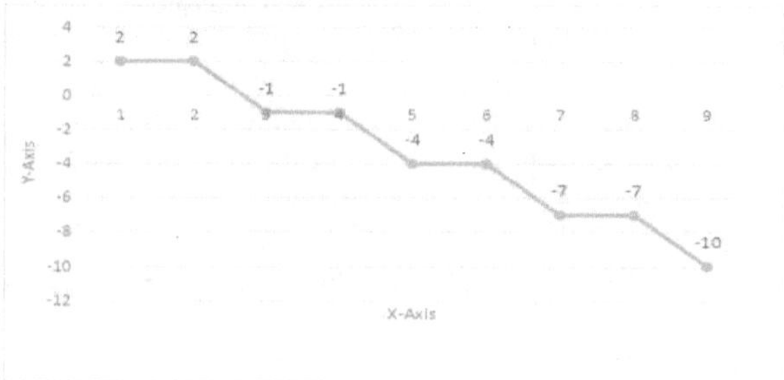

Graph 4.1.1.1: Graph of the function $y =- 3f(x) + 2$

To illustrate the necessity of using consecutive points, consider the following example based on the current function $y =- 3f(x) + 2$. Refer to Siyavula and Volunteers (2012) for a detailed explanation of the gradient method.

x	1	2	3	4	5	6	7
$f(x)$	0	0	1	1	2	2	3
y	2	2	-1	-1	-4	-4	-7

Table 4.1.1.1: Table of values for y

From this table, we compute:

$$m_1 = \frac{-4-(-1)}{5-4} =- 3$$

and

$$m_2 = \frac{-4-(-4)}{6-5} = 0$$

The presence of both gradients confirms that the function exhibits alternating slopes. Crucially, the use of consecutive x-values ensures that the corresponding points are also consecutive, which is essential for accurately determining specific gradients using the gradient method.

- Let us now examine two gradient components more directly, as previously introduced. Consider the general form of an irregular function:

$$y = d_1 f(x) \pm d_2 g(x) \pm a$$

For illustration, we focus on the specific case:

$$y = w(f(x), g(x)) = 5f(x) + g(x) + 1$$

To compute the derivatives of this function, we employ **partial differentiation**, which involves taking derivatives with respect to one control function while treating the other as constant. Unlike traditional calculus, where differentiation is performed with respect to x, here we differentiate with respect to $f(x)$ and $g(x)$ as mentioned before.

For foundational understanding of partial derivatives, refer to Stewart (2008) or any other relevant calculus textbook.

Applying partial differentiation:

- With respect to $f(x)$:

$$D_1 w = \frac{\partial w}{\partial f(x)} = 5$$

- With respect to g(x):

$$D_2 w = \frac{\partial w}{\partial g(x)} = 1$$

In each case, the other control function is treated as constant. To validate these results, we apply the **first principles** definition of the derivative.

Derivative with Respect to $f(x)$ via First Principles:

$$D_1 w = \lim_{h \to 0} \frac{w\big(f(x) + h, g(x)\big) - w(f(x), g(x))}{h}$$

Substituting the function:

$$= \lim_{h \to 0} \frac{5(f(x) + h) + g(x) + 1 - (5f(x) + g(x) + 1)}{h}$$

Simplifying:

$$D_1 w = \lim_{h \to 0} \frac{5h}{h} = 5$$

This confirms the earlier result obtained through direct differentiation. Similarly, the derivative with respect to $g(x)$ can be verified using the same method, yielding:

$$D_2 w = 1$$

For further exploration of limits and their role in differentiation, consult any of the calculus texts referenced earlier or other suitable resources.

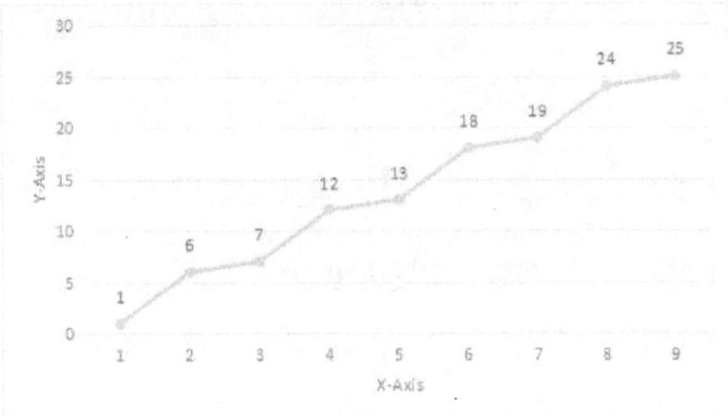

Graph 4.1.1.2: Graph of the function $y = 5f(x) + g(x) + 1$

Note that the function under consideration is a **linear irregular function**, which explains why both partial derivatives are constants. This mirrors behaviour seen in linear algebra, where linear functions yield constant gradients. The function and others that we have worked with are part of the extension of the theory of straight lines.

The presence of two distinct gradients indicates that the function is composed of two separate linear components. In the next sub-section, we will explore irregular functions composed of multiple linear segments, each with its own gradient—introducing a richer structure and more nuanced derivative behaviour.

4.1.2. Derivatives of a Summation-Based Irregular Function

- Let us now examine a function defined via a summation expression:

$$y = 1 + \sum_{i=1}^{x-1} (-1)^i (i-1)$$

This sigma function behaves as described in Section introducing irregular functions, where its irregular nature was explored. The independent variable x determines the values of the summation through partial sums, effectively acting as a control input. For compactness, we may express the function recursively as:

$$y = 1 + S_x$$
$$where\ S_x = \sum_{i=1}^{x-1} (-1)^i (i-1)$$

However, this reformulation is not particularly helpful for differentiation purposes, so we retain the original summation form when analysing gradients.

For reference on differentiating power series, consult Khan (2016) or other relevant online resources.

x	1	2	3	4	5	6	7	8
S_x	0	0	1	-1	2	-2	3	-3
y	1	1	2	0	3	-1	4	-2

Table 4.1.2.1: Table of values showing the summation values as well

From the table, we observe that S_x functions as a control variable: it receives input from x and produces intermediate values that determine the dependent variable y. This behavior is characteristic of control functions, which may be viewed as pre-dependent or post-independent variables.

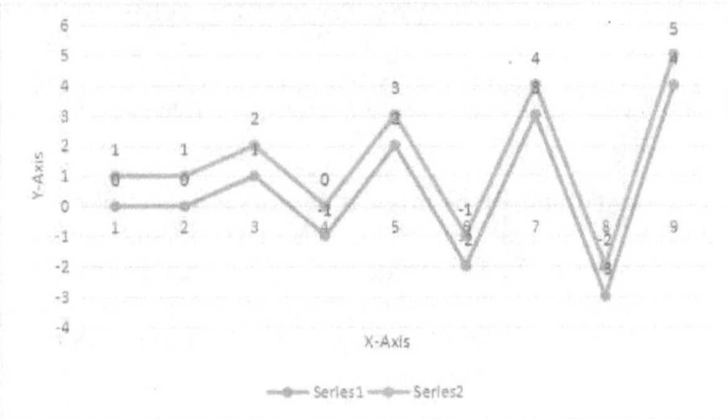

Graph 4.1.2.1: Graph of the current summation function

In the graph, the orange curve represents the function y, while the blue curve corresponds to S_x. The function is composed of multiple linear segments, each with its own gradient. These gradients can be computed between any two points on a given segment.

To determine the derivative of the function, we proceed as follows:

1. The derivative of the constant term 1 is zero.
2. The derivative of the summation is obtained by removing the summation symbol and retaining the summand:

$$\frac{dy}{dx} = (-1)^i(i-1)$$

This expression defines the gradient for each segment of the function. Because the summand alternates in sign, the graph consists of joined line segments with alternating slopes. Remember you can substitute i with x in the derivative equation.

Example: Gradient of the Fourth Segment

To compute the gradient of the fourth segment, substitute $i = 4$ into the gradient equation:

$$\frac{dy}{dx} = (-1)^4(4-1) = 3$$

Alternatively, using the gradient method between two consecutive points on the fourth line $P(4, 0)$ and $Q(5, 3)$:

$$\frac{dy}{dx} = \frac{y_2 - y_1}{x_2 - x_1} = \frac{3-0}{5-4} = 3$$

Or, using the full summation form:

$$\frac{\Delta y}{\Delta x} = \frac{1 + \sum_{i=1}^{5-1}(-1)^i(i-1) - \left(1 + \sum_{i=1}^{4-1}(-1)^i(i-1)\right)}{5-4}$$
$$= 3$$

This confirms that the derivative obtained from the gradient equation matches the slope of the fourth line calculated from the graph, validating our differentiation method. Each segment of the function contributes a distinct gradient, forming the following alternating sequence of gradients:

$$\{a_m\} = 0, 1, -2, 3, -4, 5, -6, \ldots$$

This sequence reflects the alternating nature of the summation and the structure of the function, which is composed of infinitely many linear segments indexed by i.

Comparing Derivative Methods: Gradient-Based vs. Power Series Approach

A natural question arises: what if we apply the traditional calculus method—specifically, the power series approach—to differentiate the summation-based functions discussed in this paper? Would the results differ significantly, and if so, does that invalidate the gradient-based method we've employed?

To explore this, consider the function:

$$y = \sum_{i=1}^{x} (-1)^i (i-1)$$

Using the power series method, the derivative is typically taken term-by-term, treating each summand as a function of x. This yields:

$$\frac{dy}{dx} = \sum_{i=1}^{x} (-1)^i$$

Now, let us evaluate both methods at x = 4:

- **Power Series Method**:

$$\frac{dy}{dx} = -1 + 1 - 1 + 1 = 0$$

- **Gradient-Based Method (as developed in this book)**:

$$\frac{dy}{dx} = (-1)^4 (4-1) = 3$$

The discrepancy is clear. The power series method yields a derivative of zero, while the gradient-based method produces a value of 3. To determine which is more appropriate, we validate using the gradient formula between two consecutive points:

$$m = \frac{3-0}{5-4} = 3$$

This confirms the accuracy of the gradient-based method for functions composed of joint linear segments. The derivative obtained aligns with the actual slope of the segment, reinforcing the validity of our approach.

However, the method does not universally replicate the behaviour of the power series derivative. For instance, consider:

$$y = \sum_{i=1}^{x} 4$$

This summation simplifies to y = 4x, a standard linear function. The gradient-based method yields:

$$\frac{dy}{dx} = 4$$

Yet, if we apply our earlier gradient-based summation derivative rule, we do not obtain zero—contrary to what the power series method would suggest for a constant summand. This highlights a limitation of the power series approach when applied to purely constant summations as the one above.

Interestingly, for other functions such as:

$$y = \sum_{i=1}^{x} 2i$$

both methods yield consistent results:

- **Power Series Method**:

$$\frac{dy}{dx} = \sum_{i=1}^{x} 2 = 2x$$

- **Gradient-Based Method**:

$$\frac{dy}{dx} = 2x$$

Evaluating at x = 3, both methods produce:

$$\frac{dy}{dx} = 6$$

This agreement suggests that for certain classes of functions—particularly those with linear growth—the two methods converge.

Conclusion

The functions that we have analysed in this book up to so far are either linear or composed of multiple joined linear segments. As such, their derivatives behave analogously to those of straight lines. In these cases, the **common difference** serves as both the gradient and the derivative just as you would expect a straight line to behave. The gradient-based method proves to be robust and accurate for this class of functions, especially when validated against the rise-over-run formula.

However, caution is warranted when extending this method to functions with constant summands or non-linear behaviour. The power series method remains essential for broader classes of functions, and its limitations in this context highlight the need for tailored approaches when dealing with irregular structures.

- Consider the function defined by

$$y = -1 + \sum_{i=1}^{x-1} \frac{(-1)^{i-1}(4f(i) + g(i) + 5)}{2}$$

Using the differentiation approach outlined in this paper, we obtain the derivative:

$$\frac{dy}{dx} = \frac{(-1)^{x-1}(4f(x) + g(x) + 5)}{2}$$

In contrast, applying conventional calculus techniques for differentiating series presents complications due to the presence of two discrete functions, $f(i)$ and $g(i)$, within the summation. Partial differentiation may offer a workaround, yet empirical evidence suggests that alternating summation functions are more effectively handled using the method proposed herein.
To illustrate this, we construct a table of values for the function and visualize its behavior:

x	1	2	3	4	5	6	7	8
$f(i)$	0	1	1	2	2	3	3	4
$g(i)$	0	0	1	1	2	2	3	3
S_x	0	2.5	-2	3	-4	3.5	-6	4
y	-1	1.5	-3	2	-5	2.5	-7	3

Table 4.1.2.2: Table of values for y

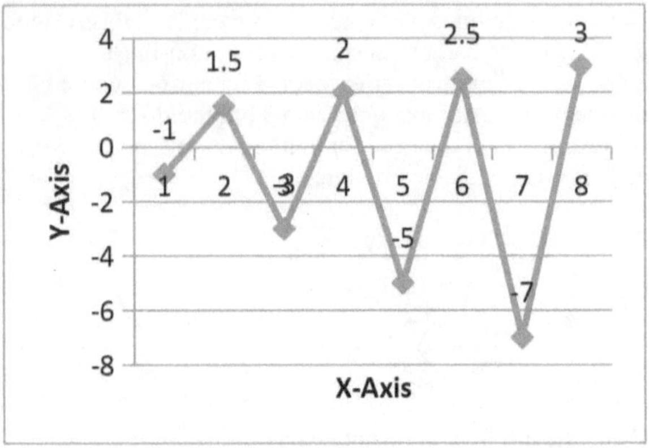

Graph 4.1.2.2: The graph of the summation function

Let us now compute the gradient of the fourth segment. Given the steep descent, the slope is expected to be negative. Using the standard gradient formula between the points (4, 2) and (5, -5), we find:

$$m = \frac{y_2 - y_1}{x_2 - x_1} = \frac{-5 - 2}{5 - 4} = -7$$

Applying the derivative expression derived earlier for $x = 4$, we obtain:

$$\frac{dy}{dx} = \frac{(-1)^{x-1}(4f(x) + g(x) + 5)}{2}$$

$$= \frac{(-1)^3(4 \times 2 + 1 + 5)}{2} = -\frac{14}{2}$$

$$= -7$$

This result aligns precisely with the gradient calculated via the rise-over-run method, reinforcing the validity of our differentiation approach.
By contrast, employing the power series method yields a partial derivative with respect to $f(i)$:

$$\frac{dy}{df(x)} = \sum_{i=1}^{x-1} (-1)^{i-1}(2)$$

For $i = 4$, this becomes:

$$\frac{dy}{df(x)} = \sum_{i=1}^{3} (-1)^{i-1}(2) = 2 - 2 + 2 = 2$$

This outcome diverges from the gradient of the fourth segment, indicating that the power series method does not faithfully capture the behaviour of the function in this context. Consequently, we affirm that the differentiation technique developed in this book offers a more accurate and pedagogically sound framework for handling summation-defined functions.

- We now examine a function defined by a nested summation structure:

$$y = 6 + \sum_{i=1}^{x-1} -3 + \sum_{i=1}^{x-1}\sum_{j=1}^{i-1} (-1)^{j-1}(3j + 7)$$

By factoring the repeated inner sum, the expression simplifies to:

$$y = 6 + \sum_{i=1}^{x-1} (-3 + \sum_{j=1}^{i-1} (-1)^{j-1}(3j + 7))$$

First Derivative
Analogous to our earlier treatment of single summation functions, we differentiate term by term to obtain the first derivative:

$$\frac{dy}{dx} =- 3 + \sum_{j=1}^{i-1} (-1)^{j-1}(3j + 7)$$

Second Derivative
Proceeding to the second derivative, we differentiate the summation with respect to x, treating i as the upper bound and recognizing the discrete nature of the index:

$$\frac{d^2y}{dx^2} = (-1)^{j-1}(3j + 7)$$

Although the indexing variables are not explicitly redefined at each stage, the intent is to guide the reader through the gradient behaviour of the function. The alternating sign in the second derivative introduces ambiguity in sign interpretation, which we resolve by considering the absolute value:

$$|\frac{d^2y}{dx^2}| = |(-1)^{j-1}(3j + 7)|$$

$$|\frac{d^2y}{dx^2}| = 3j + 7$$

Third Derivative

Differentiating the absolute second derivative yields a constant third derivative:

$$\frac{d^3y}{dx^3} = 3$$

This result is consistent with the linear growth observed in the sequence of absolute second derivative values.

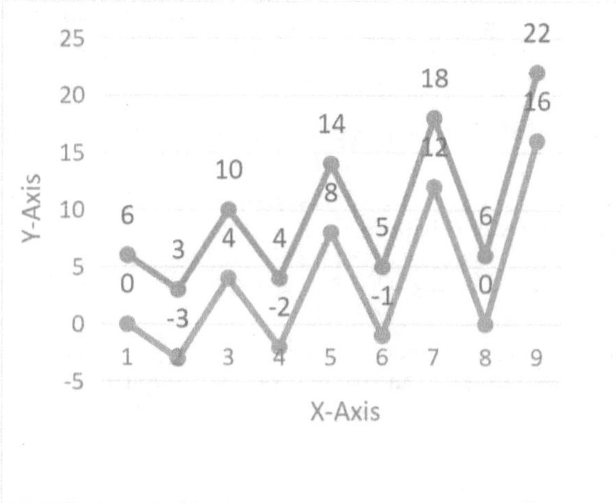

Graph 4.1.2.3: Illustration of the behaviour of the summation function.

The blue curve represents the original summation function, while the second curve—shifted vertically up by 6 units—reflects the addition of the constant term. This vertical translation will be addressed in the subsequent section on transformations.

Reconstructing the Original Function from Its Derivatives

To reverse-engineer the original function from its third derivative, we begin by evaluating the absolute second derivative at $j = 1$:

$$\left|\frac{d^2y}{dx^2}\right| = 3(1) + 7 = 10$$

This yields the first term of the positive second derivative sequence. Continuing with successive natural numbers, we obtain:

$$10, 13, 16, 19, 22, 25, \ldots$$

This arithmetic sequence is governed by the expression $3j + 7$. Reintroducing the alternating sign pattern, with the first term positive, we recover the second derivative:

$$(-1)^{j-1}(3j + 7)$$

To derive the first derivative, we sum this alternating sequence and include the constant term -3:

$$-3 + \sum_{j=1}^{i-1} (-1)^{j-1}(3j + 7)$$

Finally, to reconstruct the original function y, we sum the first derivative and add the initial constant term 6:

$$y = 6 + \sum_{i=1}^{x-1} \left(-3 + \sum_{j=1}^{i-1} (-1)^{j-1}(3j + 7) \right)$$

This expression completes the derivation of the main sequence, which exhibits the values:

$$6, 3, 10, 4, 14, 5, \ldots$$

4.1.3. Partial Derivatives and Gradient Analysis of a Composite Function

- Consider the function defined by:

$$y = 2(g(x) - 3) + f(x)(2g(x) - 1)$$

Expanding and simplifying yields:

$$y = 2g(x) - 6 + 2f(x)g(x) - f(x)$$

To analyse the behaviour of this function, we compute its partial derivatives with respect to $f(x)$ and $g(x)$:

- With respect to $f(x)$:

$$\frac{dy}{df(x)} = 2g(x) - 1$$

- With respect to $g(x)$:

$$\frac{dy}{dg(x)} = 2 + 2f(x)$$

Both expressions are correct and provide insight into how changes in each variable affect the output of the function. To visualize this, we construct a table of values:

x	1	2	3	4	5	6	7	8	9
$f(x)$	0	1	1	2	2	3	3	4	4
$g(x)$	0	0	1	1	2	2	3	3	4
y	-6	-7	-3	-2	4	7	15	20	30

Table 4.1.3.1: Table of values for $y = 2g(x) - 6 + 2f(x)g(x) - f(x)$

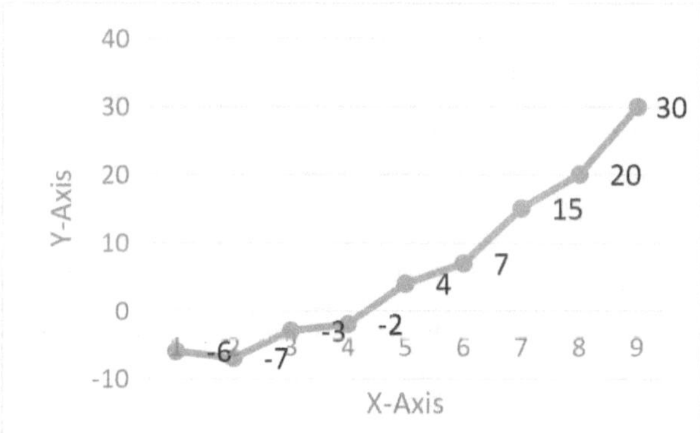

Graph 4.1.3.1: Illustrates the function and its piecewise linear segments, each corresponding to a distinct gradient.

Gradient Verification via Derivatives and Finite Differences

To validate the partial derivatives, we examine the gradient of individual segments. For the fifth segment (between

$x = 5$ and $x = 6$), we apply the derivative with respect to $f(x)$:

- From the table, $g(5) = 2$, so:

$$\frac{dy}{df(x)} = 2g(5) - 1 = 2 \times 2 - 1 = 3$$

Using the finite difference method:

$$m = \frac{y_2 - y_1}{x_2 - x_1} = \frac{7 - 4}{6 - 5} = 3$$

The results are consistent, confirming the validity of the derivative.
Similarly, for the sixth segment (between $x = 6$ and $x = 7$), we use the derivative with respect to $g(x)$:

- From the table, $f(6) = 3$, so:

$$\frac{dy}{dg(x)} = 2 + 2f(6) = 2 + 2(3) = 8$$

Using the gradient formula:

$$m = \frac{y_2 - y_1}{x_2 - x_1} = \frac{15 - 7}{7 - 6} = 8$$

Again, the derivative matches the observed gradient.

Structural Insight: Piecewise Composition and Derivative Assignment

This function, like many irregular functions studied herein, is composed of multiple linear segments joined to form a continuous whole. Each segment corresponds to a distinct derivative expression, depending on its position in the sequence:

- Odd-numbered segments (e.g., 1st, 3rd, 5th, etc.) are governed by:

$$\frac{dy}{df(x)} = 2g(x) - 1$$

- Even-numbered segments (e.g., 2nd, 4th, 6th, etc.) are governed by:

$$\frac{dy}{dg(x)} = 2 + 2f(x)$$

This alternating structure reflects the composite nature of the function and underscores the importance of identifying the appropriate derivative expression for each segment under analysis.

4.2. Derivatives and Limits in the Context of Discrete and Continuous Functions

4.2.1. Analysis of Gradients

This section has been published in the preprint article *The Extension of the Theory of Straight Lines and Second Degree Irregular Functions* by (Mbele, 2025). It is important to recognize that the derivatives of quadratic functions, particularly those derived from summation-defined sequences, do not always yield constant common differences as one might expect from linear functions. This observation leads us to a fundamental question: if the two expressions

- $y = \sum_{i=1}^{x} 2i$
- $y = x^2 + x$

generate the same sequence of values, why do their derivatives differ?

The derivative of the summation function yields:

$$\frac{dy}{dx} = 2x,$$

whereas the derivative of the quadratic function is:

$$\frac{dy}{dx} = 2x + 1.$$

This discrepancy prompts a deeper investigation. In linear functions, the derivative often coincides with both the gradient and the common difference of the function or sequence. For example, consider the linear function:

$$y = 3x + 4$$

Its derivative is:

$$\frac{dy}{dx} = 3$$

and the sequence it generates—$4, 7, 10, 13, \ldots$—has a constant common difference of 3. The gradient between any two successive points on the graph of this function is:

$$m = \frac{y_2 - y_1}{x_2 - x_1} = \frac{10 - 7}{3 - 2} = 3,$$

confirming that the derivative, gradient, and common difference are all equivalent in this linear case.

Secant vs. Tangent Gradients: A Conceptual Distinction

Returning to the summation function $y = \sum_{i=1}^{x} 2i$, which produces the sequence $2, 6, 12, 20, 30, 42, 56, \ldots$ we observe that its derivative $\frac{dy}{dx} = 2x$ corresponds to the gradient of secant lines—that is, the average rate of change between successive terms. In contrast, the derivative of the equivalent quadratic expression $y = x^2 + x$ (which generates the same sequence of values) yields $\frac{dy}{dx} = 2x + 1$, which represents the gradient of tangent lines—the instantaneous rate of change at each point.

This distinction clarifies why the derivatives differ despite the functions generating identical sequences. The summation-based derivative captures discrete transitions (secant behaviour), while the polynomial derivative reflects continuous variation (tangent behaviour).

This conceptual nuance has not been explicitly highlighted in prior works, including those by Mbele (2025). To formalize this distinction, we propose the following:

Secant Gradient Representation: For summation-defined functions, the gradient between terms can be denoted by

$$\frac{dy}{dx} = 2x,$$

- representing the changing common difference.

Tangent Gradient Representation: For continuous polynomial functions defining the same sequence, the derivative

$$\frac{dy}{dx} = 2x + 1$$

- captures the instantaneous rate of change,

This clarification invites a broader reclassification of the study presented herein—and in previous works by Mbele (2024)—as an investigation of linear and nonlinear behaviours across both discrete and continuous domains. While the previous focus areas were on straight-line behaviour, subsequent sections on chapter 5 will extend the analysis to curved functions.

Derivative Extraction from Summation Functions

As established in earlier articles (Mbele, 2025) or in section 4.1 of this book, the derivative of a summation function is obtained by removing the sigma notation and retaining the summand. For example:

If

$$y = \sum_{i=1}^{x} 2i$$

then

$$\frac{dy}{dx} = 2x.$$

This method reflects the discrete nature of the function and aligns with the secant interpretation of gradients (behaves like a linear function).

Tangent vs. Secant Gradients: Derivatives and Common Differences in Parabolic Functions

Let us now consider the function:

$$y = x^2,$$

whose derivative is:

$$\frac{dy}{dx} = 2x.$$

This derivative, an even function, represents the gradient of the tangent line at any given point on the curve. However, it is not equivalent to the common difference of the sequence generated by this function.

For instance, the sequence of values produced by $y = x^2$ is:

$$1, 4, 9, 16, 25, 36, 49, \dots$$

The successive differences between terms—i.e., the common differences—form the sequence:

$$1, 3, 5, 7, 9, 11, 13, \dots$$

This sequence is defined by the linear expression (1 has been inserted and is the term that comes before 3):

$$\Delta y = 2x - 1,$$

which clearly differs from the derivative $\frac{dy}{dx} = 2x$. Thus, in this context, the derivative does not coincide with the common difference.

General Principle: When Is the Derivative Also the Common Difference?

We conclude that:

- For linear functions (or functions composed of multiple linear segments), the derivative is simultaneously the gradient and the common difference.

- For nonlinear functions, such as parabolas, the derivative represents the tangent gradient, while the common difference corresponds to the secant gradient—the average rate of change between successive terms.

This distinction is critical when analysing discrete sequences derived from continuous functions.

Example: Secant Gradient in a Parabolic Function

Consider the secant line passing through the points $(2, 4)$ and $(3, 9)$ on the curve $y = x^2$. The gradient is:

$$m_{sec} = \frac{9-4}{3-2} = 5.$$

This matches the value obtained from the common difference expression:

At $x = 3$:

$$\Delta y = 2x - 1 = 2(3) - 1 = 5$$

Comparison with a Parabolic Series Function

Now consider the summation-defined function:

$$y = \sum_{i=1}^{x} 2i,$$

which produces the sequence:

$$2, 6, 12, 20, 30, 42, 56, \ldots$$

This function is equivalent to the quadratic expression:

$$y = x^2 + x,$$

and its derivative is:

$$\frac{dy}{dx} = 2x + 1.$$

However, when we differentiate the summation directly—by removing the sigma notation—we obtain:

$$\frac{dy}{dx} = 2x,$$

which, as discussed, corresponds to the gradient of the secant line rather than the tangent.

To verify this, consider the points (2, 6) and (3, 12):

- Secant gradient:

$$m_{sec} = \frac{y_2 - y_1}{x_2 - x_1} = \frac{12 - 6}{3 - 2} = 6.$$

- Using the derivative:

$\frac{dy}{dx} = 2x$ and at $x = 3$ we obtain $\frac{dy}{dx} = 2(3) = 6$

Both methods yield consistent results, confirming that the derivative $2x$ reflects the secant gradient in this context (not the tangent gradient).

Limit-Based Verification

To further validate the secant gradient, we apply the limit definition:

$$m_{sec} = \lim_{x \to 2} \frac{\sum_{i=1}^{3} 2i - \sum_{i=1}^{2} 2i}{3 - 2}$$

$$= \lim_{x \to 2} \frac{(2 + 4 + 6) - (2 + 4)}{1}$$

$$\therefore \ m_{sec} = 6$$

This result aligns with previous calculations and reinforces the interpretation of the derivative as a secant gradient in summation-defined functions.

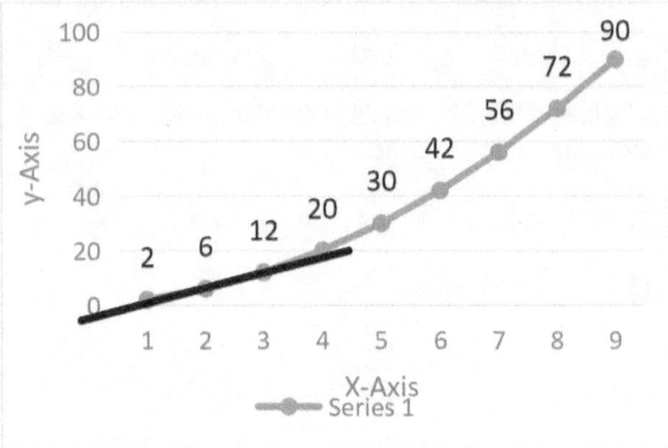

Graph 4.2.1.1: Illustration of the secant line passing through the points $(2, 6)$ and $(3, 12)$, visually confirming the gradient

---

Derivatives via Limits: Tangent vs. Secant Gradients in Summation-Defined Functions

To derive the gradient of the tangent line for the function

$$y = x^2 + x,$$

—which is algebraically equivalent to the summation-defined function

$$y = \sum_{i=1}^{x} 2i$$

(see Chapter 5, Section 5.1)—we apply the limit definition of the derivative:

$$f'(x) = \lim_{h \to 0} \frac{\sum_{i=1}^{x+h} 2i - \sum_{i=1}^{x} 2i}{h}$$

Using the identity

$$\sum_{i=1}^{x} i = \frac{x(x+1)}{2},$$

we rewrite the expression as:

$$f'(x) = \lim_{h\to 0} \frac{2\left(\frac{(x+h)(x+h+1)}{2}\right) - 2(\frac{x(x+1)}{2})}{h}$$

Cancelling the factor of 2 and expanding:

$$= \lim_{h\to 0} \frac{x^2 + 2xh + h^2 + x + h - (x^2 + x)}{h}$$

$$= \lim_{h\to 0} \frac{h(2x+h+1)}{h}$$

Simplifying:

$$f'(x) = 2x + 1.$$

This derivative represents the gradient of the tangent line at a specific point on the curve, denoted m_{tan}. For example, at $x = 7$, we compute:

$$f'(7) = 2(7) + 1 = 15\,.$$

Verifying the Tangent Gradient via Limit Evaluation

To confirm the result example above using the limit definition directly, we evaluate:

$$f'(x) = \lim_{x\to 7} \frac{\sum_{i=1}^{x} 2i - \sum_{i=1}^{7} 2i}{x-7}$$

Given:

$$\sum_{i=1}^{7} 2i = 2 + 4 + 6 + 8 + 10 + 12 + 14 = 56$$

we substitute:

$$f'(x) = \lim_{x\to 7} \frac{2\left(\frac{x(x+1)}{2}\right) - 2(\frac{7(7+1)}{2})}{x-7}$$

$$= \lim_{x\to 7} \frac{x^2 + x - 56}{x-7}.$$

Factoring the numerator:

$$= \lim_{x\to 7} \frac{(x-7)(x+8)}{x-7},$$

and cancelling:

$$= \lim_{x\to 7} (x+8) = 7 + 8 = 15.$$

This confirms that the derivative obtained via the limit method matches the result from the power rule, validating both approaches.

Conceptual Conclusion: Derivatives, Gradients, and Common Differences

We conclude that:

- For curved functions such as $y = x^2 + x$, the derivative yields the tangent gradient m_{tan}, not the common difference.

- The common differences in such functions correspond to the secant gradients, computed between successive terms.

- For linear functions or functions composed of multiple linear segments, the derivative simultaneously represents both the gradient and the common difference. Each segment inherits its gradient directly from the derivative.

This distinction is essential when interpreting the behaviour of summation-defined functions and their polynomial equivalents (See section 5.1). The derivative provides instantaneous change (tangent), while common differences reflect average change (secant) across discrete intervals.

Derivatives in Irregular Functions: further Secant vs. Tangent Interpretations

To further clarify the distinction between derivatives as gradients and derivatives as common differences, we examine a new (another) summation-defined function:

$$y = \sum_{i=1}^{x-1} (-1)^i (i+1).$$

This function generates the sequence:

$$0, -2, 1, -3, 2, -4, 3, -5, 4, \ldots$$

Computing the common differences between successive terms yields:

$$-2, 3, -4, 5, -6, 7, -8, \ldots$$

Investigative Aim

Our goal is to determine whether the derivative of this function behaves like that of a linear function—where the derivative equals both the gradient and the common difference—or like that of a curved function, such as a parabola, where the derivative represents only the tangent gradient.

Graph 4.2.1.2 illustrates the structure of the function. It is composed of multiple straight-line segments joined together, forming a piecewise linear graph. Each segment possesses its own distinct gradient.

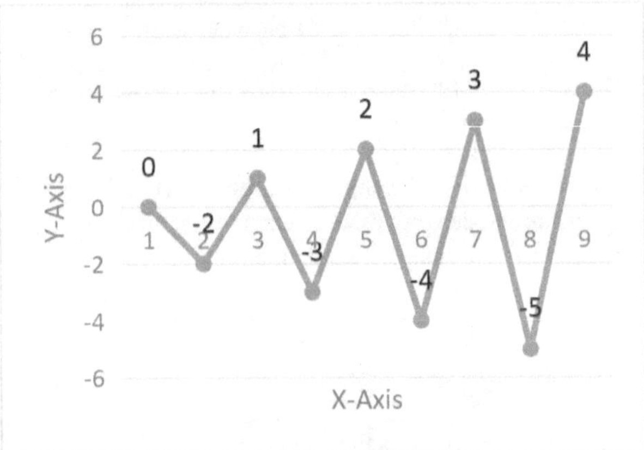

Graph 4.2.1.2: Graph of the summation function

Derivative as Common Difference and Gradient

Given the linear nature of each segment, the derivative of this function corresponds to both the gradient and the common difference. Importantly, the appropriate derivative method here is the secant-based approach, where we remove the summation symbol and differentiate the summand directly:

$$\frac{dy}{dx} = (-1)^x(x+1).$$

This expression captures the alternating structure and linear growth of the function. For example, at $x = 3$:

$$\frac{dy}{dx} = (-1)^3(3+1) =-4,$$

which matches the third common difference:

$$CD = y_4 - y_3 =-3-1 =-4.$$

Thus, in this context, the derivative serves as both the gradient and the common difference. The kinds of functions displaying these characteristics are part of the extension of the theory of straight lines.

Conceptual Clarification

This behaviour contrasts with that of smooth curves, such as $y = x^2 + x$, where the derivative (e.g., $2x + 1$) represents the tangent gradient but not the common difference. In such cases, the common differences correspond to secant gradients and must be computed separately.

For the current function, however, the derivative derived from the summand directly reflects the discrete transitions between terms. Attempting to apply the limit-based derivative method (as used for smooth curves) would not yield meaningful tangent gradients, since the function lacks continuous curvature.

Terminological Proposal

To avoid confusion, we propose distinguishing between two derivative types:

- Secant-Derivative: Derived by removing the summation symbol and remaining with the summand. Applicable to discrete or piecewise-linear functions. Represents both gradient and common difference.

- Tangent-Derivative: Derived via limit-based methods or power rules. Applicable to smooth, continuous functions. Represents the instantaneous rate of change (tangent gradient), not the common difference.

This clarification addresses assumptions made in the earlier preprint article *Irregular Functions: An Introduction* (Mbele, 2024), and will inform our on-going exploration of derivative behaviour in irregular and hybrid functions.

Chapter 5

Information from this chapter has also been published in the preprint article *The Extension of the Theory of Straight Lines and Second Degree Irregular Functions* by (Mbele, 2025).

5.1 Quadratic Nature of Summation Functions

We begin by examining whether the summation functions $y = \sum_{i=1}^{x} i$ and $y = \sum_{i=1}^{x} 2i$ exhibit quadratic behaviour or not.

5.1.1 Analysis of the First Summation Function: $y = \sum_{i=1}^{x} i$

To investigate the nature of this function, we compute partial sums by substituting successive integer values for the independent variable x, thereby generating corresponding values for the dependent variable y (we will denote y as T_x in order to work with the quadratic sequences):

- $x = 1$ then $y = 1$
- $x = 2$ then $y = 1 + 2 = 3$
- $x = 3$ then $y = 1 + 2 + 3 = 6$
- $x = 4$ then $y = 1 + 2 + 3 + 4 = 10$

Continuing in this manner yields the sequence:

$$1, 3, 6, 10, 15, 21, 28, \dots$$

To determine whether this sequence follows a quadratic pattern, we examine its first-order differences:

$$\{d_1\} = 2, 3, 4, 5, 6, 7, \dots$$

and then the second-order differences:

$$\{d_2\} = 1, 1, 1, 1, 1, \dots$$

The constancy of the second-order differences confirms that the sequence is quadratic in nature (the second order differences are constant). Consequently, the summation function can be expressed as a quadratic equation. To derive this equation, we solve the following system of equations based on the general quadratic form $T_x = ax^2 + bx + c$:

$$x = 1: T_1 = a + b + c = 1$$

$$x = 2: T_2 = 4a + 2b + c = 3$$

$$x = 3: T_3 = 9a + 3b + c = 6$$

Subtracting equations to eliminate c:

$$T_2 - T_1 = 4a + 2b + c - (a + b + c) = 3 - 1$$

$$3a + b = 2 \qquad ...(1)$$

$$T_3 - T_2 = 9a + 3b + c - (4a + 2b + c) = 6 - 3$$

$$5a + b = 3 \qquad ...(2)$$

Subtracting equation (1) from equation (2):

$$(2) - (1): 3 - 2 = 5a + b - (3a + b)$$

$$1 = 2a$$

$$a = \frac{1}{2} = 0.5$$

Substituting into equation (1):

$$3(0.5) + b = 2$$

$$\frac{3}{2} + b = 2$$

$$b = \frac{1}{2}$$

Using the initial condition $a + b + c = 1$:

$$0.5 + 0.5 + c = 1$$

$$c = 0$$

Thus, the quadratic equation representing the summation function is:

$$T_x = \frac{1}{2}x^2 + \frac{1}{2}x$$

Verification for $x = 4$:

$$T_4 = \frac{1}{2}(4)^2 + \frac{1}{2}(4) = 8 + 2 = 10$$

This confirms the correctness of the derived expression. Therefore, although the original summation function is defined by taking partial sums, it admits an equivalent closed-form quadratic representation and the summation function equation is not unique because we can obtain an equivalent algebraic quadratic equation for it.

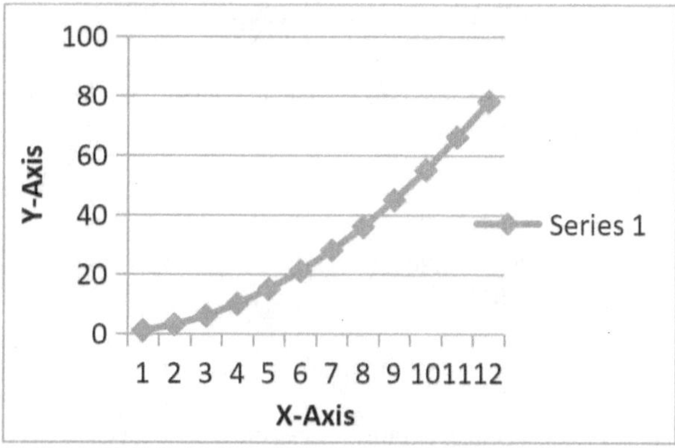

Graph 5.1.1.1: Graph of the quadratic function

5.1.2 Analysis of the Second Summation Function: $y = \sum_{i=1}^{x} 2i$

We now examine the summation function defined by $y = \sum_{i=1}^{x} 2i$, which represents the sum of the first x even natural numbers. As before, we compute partial sums to generate the sequence of values for y:

- $x = 1$ then $y = 2$
- $x = 2$ then $y = 2 + 4 = 6$
- $x = 3$ then $y = 2 + 4 + 6 = 12$
- $x = 4$ then $y = 2 + 4 + 6 + 8 = 20.$

This yields the sequence:

$$2, 6, 12, 20, 30, 42, 56, \ldots.$$

First Differences and Gradient Analysis

To assess whether this sequence is quadratic, we examine the first-order differences:

$$d_1 = 4, 6, 8, 10, 12, 14, \ldots$$

These differences increase linearly, suggesting a constant second-order difference:

$$d_2 = 2, 2, 2, 2, 2, 2, \ldots$$

This confirms the quadratic nature of the sequence. Moreover, the first differences correspond to the gradients of secant lines between successive points. For example:

$$\frac{dy}{dx} = \frac{y_2 - y_1}{x_2 - x_1} = \frac{6 - 2}{2 - 1} = 4$$

$$and\ \frac{dy}{dx} = \frac{y_3 - y_2}{x_3 - x_2} = \frac{12 - 6}{3 - 2} = 6$$

These gradients align with the expression $\frac{dy}{dx} = 2x$, which is the derivative of a summation function. The second derivative is constant:

$$\frac{dy^2}{x^2x} = 2$$

It corresponds to the second common differences we found when assessing whether the summation function is quadratic or not.

Deriving the Quadratic Equation

To find the explicit quadratic form $T_x = ax^2 + bx + c$, we solve the following system:

$$x = 1: T_1 = a + b + c = 2$$

$$x = 2: T_2 = 4a + 2b + c = 6$$

$$x = 3: T_3 = 9a + 3b + c = 12$$

Subtracting equations:

$$T_2 - T_1 = 4a + 2b + c - (a + b + c) = 6 - 2$$

$$4 = 3a + b \qquad ... (1)$$

$$T_3 - T_2 = 9a + 3b + c - (4a + 2b + c) = 12 - 6$$

$$6 = 5a + b \qquad ... (2)$$

Subtracting (1) from (2):

$$(2) - (1) = 5a + b - (3a + b) = 6 - 4$$

$$2 = 2a$$

$$a = \frac{2}{2} = 1$$

Substituting into equation (1):

$$3(1) + b = 4$$

$$3 + b = 4$$

$$b = 1$$

Using the initial condition:

$$a + b + c = 2$$

$$1 + 1 + c = 2$$

$$c = 0$$

Thus, the quadratic equation is:

$$y = x^2 + x$$

Verification for $x = 5$:

$$y = 5^2 + 5 = 25 + 5 = 30$$

This matches the fifth term in the sequence, confirming the correctness of the derived expression.

Conclusion

The summation function $y = \sum_{i=1}^{x} 2i$ is quadratic because it sums a first-degree linear term (i.e. $2x$). More generally, any summation function of the form $y = \sum_{i=1}^{x} (ai + b)$ will yield a quadratic expression in x, due to the accumulation of linear growth.

5.2 Irregular Second-Degree Functions

We now explore a class of second-degree functions that exhibit irregular behaviour due to the structure of their input variable.

5.2.1. The curvy-straight function with diminishing g(x) term

Consider the function:

$$y = 3f^2(x) + 2 \qquad when \qquad |x| = 2$$

Here, the notation $|x| = 2$ refers not to the absolute value in the traditional sense, but to the cardinality or repetition count—indicating that each value of $f(x)$ is repeated twice before incrementing. This creates a stepwise structure in the input function $f(x)$, which in turn affects the behaviour of y.

Tabulated Values

The following table illustrates the values of the pre-independent variable x, the post-independent variable $f(x)$, its square $f^2(x)$, and the resulting dependent variable y:

x	1	2	3	4	5	6	7	8	9
$f(x)$	0	0	1	1	2	2	3	3	4
$f^2(x)$	0	0	1	1	4	4	9	9	16
y	2	2	5	5	14	14	29	29	50

Table 5.2.1.1: Values of x, $f(x)$, $f^2(x)$, and y

Derivative Analysis

The function $y = 3f^2(x) + 2$ is a composite function, and its derivative with respect to $f(x)$ is obtained using the power rule:

$$\frac{dy}{dx} = \frac{d}{dx}\left(3f^2(x) + 2\right) = 6f(x)$$

However, since this function originates from a broader family of expressions of the form:

$$y = d_1 f^2(x) \pm d_2 g^2(x) \pm a$$

we must also consider the derivative with respect to $g(x)$. In this case, the $g(x)$ term is absent (or constant), so:

$$\frac{dy}{dg(x)} = 0$$

This indicates that $g(x)$ does not contribute to the rate of change of y, and its influence is effectively nullified in this instance.

Local Behaviour and Shape Analysis

To understand the shape of the function, we examine its behaviour between consecutive values of x. Because $f(x)$ remains constant over intervals of two units (due to the cardinality condition), the function alternates between horizontal segments and curved transitions.

- Between $x = 1$ and $x = 2$, $f(x) = 0$, so $y = 2$ remains constant.

- Between $x = 2$ and $x = 3$, $f(x)$ increases from 0 to 1, producing a smooth curve in y.

To illustrate this, we interpolate values between $x = 2$ and $x = 3$:

x	$f(x)$	y
2	0	2
2.2	0.2	2.12
2.4	0.4	2.48
2.6	0.6	3.08
2.8	0.8	3.92
2.9	0.9	4.43
3	1	5

Table 5.2.1.2: Interpolated values between $x = 2$ and $x = 3$

This pattern—alternating between flat segments and curved transitions—repeats across the domain, e.g., between $x = 4$ and $x = 5$, $x = 6$ and $x = 7$, and so forth. The result is a piecewise function composed of horizontal lines interspersed with quadratic arcs.

Conclusion

The function $y = 3f^2(x) + 2$ demonstrates a hybrid structure: locally quadratic within intervals where $f(x)$ varies, and constant where $f(x)$ is held fixed. This irregularity arises from the cardinality-based definition of $f(x)$, making the overall function a staircase of quadratic segments.

Verifying Curvature via Tangent Gradients

To confirm that the function $y = 3f^2(x) + 2$ behaves as a curve between x = 2 and x = 3, we examine the gradients of tangent lines at two distinct points within this interval. If the gradients differ, the function is not linear over that segment.

Tangent Gradients at Two Points

Let us evaluate the derivative at:

- $x = 2.6$, where $f(2.6) = 0.6$
- $x = 2.9$, where $f(2.9) = 0.9$

Using the power rule derivative:

$$\frac{dy}{df(x)} = 6f(x)$$

We compute:

- At $x = 2.6$:

$$\frac{dy}{df(x)} = \frac{dy}{df(2.6)} = 6(0.6) = 3.6$$

- At $x = 2.9$:

$$\frac{dy}{df(x)} = \frac{dy}{df(2.9)} = 6(0.9) = 5.4$$

Since the gradients differ, the function is confirmed to be curved between $x = 2$ and $x = 3$. This aligns with the earlier tabulated values, which showed non-linear growth in y.

Verifying the Derivative via Limits

To reinforce the correctness of the derivative, we apply the limit definition at $f(x) = f(2.9) = 0.6$:

$$y' = \lim_{f(x)\to 0.6} \frac{3f^2(x) + 2 - (3f^2(2.6) + 2)}{f(x) - 0.6}$$

$$= \lim_{f(x)\to 0.6} \frac{3f^2(x) + 2 - 3.08}{f(x) - 0.6}$$

Since $3f^2(2.6) + 2 = 3(0.6)^2 + 2 = 3.08$, we simplify:

$$= \lim_{f(x)\to 0.6} \frac{3f^2(x) - 1.08}{f(x) - 0.6}$$

Factoring the numerator:

$$= \lim_{f(x)\to 0.6} \frac{(3f(x) + 1.8)(f(x) - 0.6)}{f(x) - 0.6}$$

$$\lim_{f(x)\to 0.6} (3f(x) + 1.8)$$

$$= 3(0.6) + 1.8 = 1.8 + 1.8 = 3.6$$

This confirms the derivative value matches the power rule result.

General Derivative via Limit Definition

We now derive the general expression for the derivative using the limit definition:

$$y' = \lim_{h \to 0} \frac{3(f(x) + h)^2 + 2 - (3f^2(x) + 2)}{h}$$

Expanding the numerator:

$$= \lim_{h \to 0} \frac{3(f^2(x) + 2f(x)h + h^2) + 2 - 3f^2(x) - 2}{h}$$

Simplifying:

$$= \lim_{h \to 0} \frac{6f(x)h + 3h^2}{h}$$

$$= \lim_{h \to 0} (6f(x) + 3h) = 6f(x)$$

Thus, the limit-based derivative confirms the power rule result.

Example: Tangent Line Equation

To find the equation of the tangent line at $x = 2.6$, where:

- $f(x) = f(2.6) = 0.6$
- $y = 3f^2(2.6) + 2 = 3(0.6)^2 + 2 = 3.08$
- Gradient $m = 3.6$

Using the point-slope form:

$$y - 3.08 = 3.6(f(x) - 0.6)$$

which implies that $y = 3.6f(x) + 0.92$

The above final equation is the equation of the tangent at $x = 2.6$ or at $f(x) = f(2.6) = 0.6$.

5.2.2 Geometric Interpretation: The Shape of the Function

The dual derivative structure of the function $y = 3f^2(x) + 2$, namely:

- $\frac{dy}{df(x)} = 6f(x)$
- $\frac{dy}{dg(x)} = 0$

reveals that the function comprises both curved and linear segments. Specifically:

- The derivative with respect to $f(x)$ governs the curved portions of the graph.
- The derivative with respect to $g(x)$, being identically zero, corresponds to horizontal line segments.

Interpreting the Derivatives

- A derivative of zero implies a constant rate of change, i.e., a horizontal tangent line. Thus, wherever $g(x)$ dominates, the function remains flat.

- A derivative of the form $6f(x)$ implies a variable rate of change, producing a continuously changing slope—characteristic of a curve.

This dual behaviour results in a piecewise structure: the function alternates between horizontal lines and smooth curves, depending on which component—$f(x)$ or $g(x)$—is active.

Dimensional Considerations

Although the function internally depends on both $f(x)$ and $g(x)$, its graphical representation is two-dimensional, plotting the independent variable x against the dependent variable y. The internal structure of $f(x)$ and $g(x)$ influences the shape, but they are not plotted directly.

Graphical Interpretation

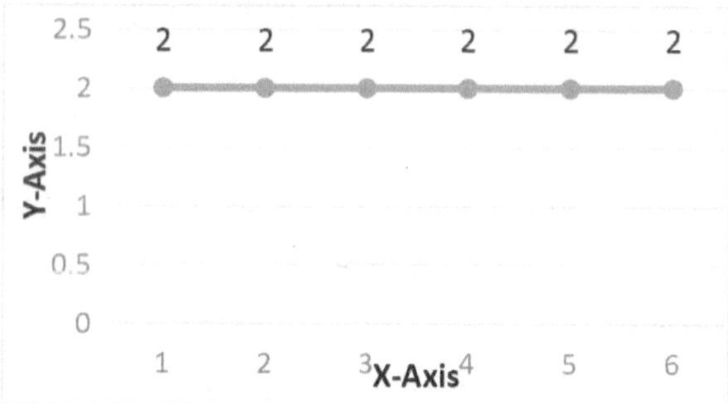

Graph 5.2.2.1: Representation of the $g(x)$ influenced portion of the graph

Represents the segment of the function between $x = 1$ and $x = 2$, where $f(x) = 0$. The graph is a horizontal line, consistent with the zero derivative from $g(x)$.

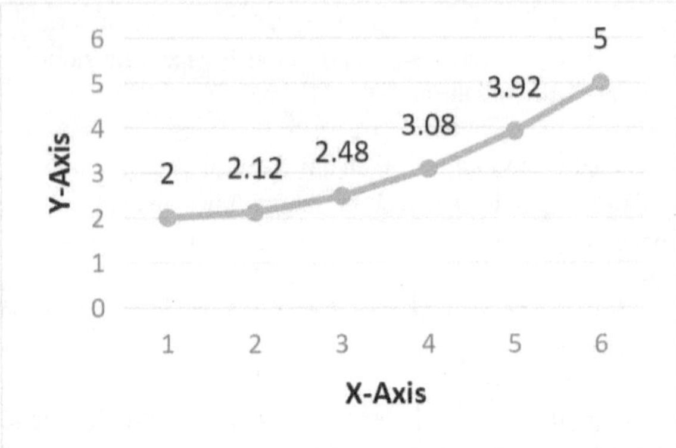

Graph 5.2.2.2: Representation of the $f(x)$ influenced portion of the graph

Depicts the segment between $x = 2$ and $x = 3$, where $f(x)$ increases from 0 to 1. The graph exhibits a smooth upward curve, confirming the derivative-based prediction.

Despite limitations in axis labelling due to the graphing utility, the visual behaviour aligns with the analytical conclusions: the function alternates between flat and curved segments, governed by the activity of $f(x)$ and $g(x)$. Beware of these limitations throughout the rest of this chapter.

Composite Representation of Function Segments

We now present a composite graph—referred to as Graph 5.2.2.3—depicting the previously analysed segments of the function. Specifically, the portion between $x = 1$ and $x = 3$, which has already been constructed, is included here. Note that the labelling along the x-axis may differ slightly from earlier figures; this should not cause confusion (later graphs will also experience the same problem..

Graph 5.2.2.3: Illustration of two parts of the function from $x = 1$ to $x = 3$

The graph illustrates two distinct segments plotted on a shared set of axes: the first segment is linear, while the second exhibits curvature. This juxtaposition exemplifies the hybrid nature of the function, which we have termed a *curvy-straight function.* The subsequent graph extends this representation to approximately four segments, incorporating both previously drawn and newly introduced components.

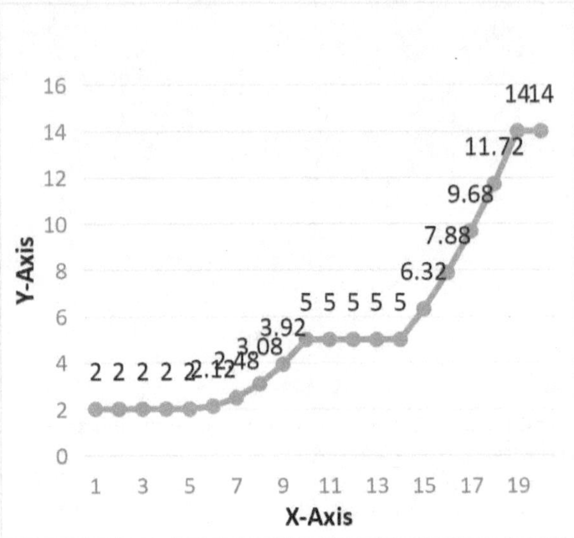

Graph 5.2.2.4: Illustrate the first four segments of the function

To contextualize this structure, consider the class of *multi-linear functions* previously discussed. These functions consist exclusively of piecewise linear segments and possess derivative functions that yield the gradients of each constituent line. In contrast, the curvy-straight function comprises a sequence of short curves and straight lines, smoothly joined to form a continuous graph. One may also encounter *multi-curved functions*, composed entirely of interconnected curves.

The key distinction lies in the composition: while multi-linear functions are constructed solely from straight segments, curvy-straight functions integrate both linear and nonlinear elements. This conceptual framework builds upon our earlier exploration in the preprint article *Irregular Functions: An Introduction* (Mbele, 2025), where the structure and behaviour of multi-linear functions were examined in detail.

5.3. Analysis of a Curvy-Straight Chain Function with Non-Diminishing Linear Component

We now examine a curvy-straight function in which the linear component $g(x)$ does not diminish, owing to the non-zero coefficient preceding it. For this analysis, we assume the cardinality of x for both control functions $f(x)$ and $g(x)$ to be 2. The function under consideration is:

$$y = 3f^2(x) + 2g(x) + 4$$

To understand the behaviour of this function, we compute its partial derivatives with respect to each control function:

- $\frac{dy}{df(x)} = 6f(x)$
- $\frac{dy}{dg(x)} = 2$

The derivative with respect to $f(x)$ varies with x, indicating a nonlinear (curved) contribution, while the derivative with respect to $g(x)$ remains constant, signifying a linear influence. This dual behavior suggests that the function is composed of distinct segments—some curved, others linear—joined together. We refer to such composite structures as **chain functions**, due to their segmented nature (or hybrid functions).

The values of $f(x)$, $g(x)$, and the resulting y are tabulated below:

x	1	2	3	4	5	6	7	8	9
$f(x)$	0	0	1	1	2	2	3	3	4
$g(x)$	0	1	1	2	2	3	3	4	4
y	4	6	9	11	20	22	37	39	60

Table 5.3.1: Values of $f(x)$, $g(x)$, and y

Interval Analysis: $x = 1$ to $x = 2$

Within the interval $x = 1$ to $x = 2$, the values of $f(x)$ remain constant at 0, implying that the quadratic term $3f^2(x)$ does not influence the function in this range. However, $g(x)$ increases from 0 to 1, and since its derivative is constant, the function behaves linearly over this interval.

To confirm this, we examine intermediate values:

x	$g(x)$	$y = 2g(x) + 4$
1	0	4
1.25	0.25	4.5
1.5	0.5	5
1.75	0.75	5.5
2	1	6

Table 5.3.2: Linear behaviour of y over $x = 1$ to $x = 2$

The consistent difference of 0.5 between successive y-values confirms the linearity of the function in this interval. This aligns with the constant gradient of 2 derived from the partial derivative with respect to $g(x)$. The function segment in this interval is governed by:

$$y = 2g(x) + 4$$

This linear sub-function represents the first **chain segment** of the overall function. It is important to note that while this expression accurately describes the behaviour between $x = 1$ and $x = 2$, similar forms (e.g., $y = 2g(x) + 4$) may appear in other intervals such as $x = 5$ to $x = 6$, though each segment is defined by its unique values of $f(x)$ and $g(x)$.

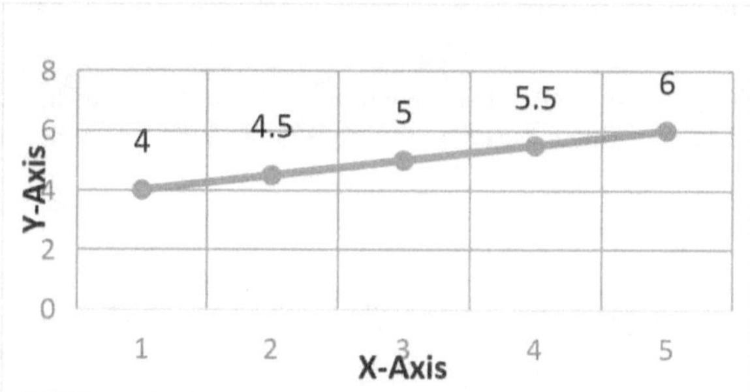

Graph 5.3.1: Linear segment $y = 2g(x) + 4$ between $x = 1$ and $x = 2$

In the next stage of analysis, we will investigate the interval $x = 2$ to $x = 3$, where both $f(x)$ and $g(x)$ begin to vary, introducing curvature into the function. This will allow us to characterize the next chain segment and further understand the composite structure of the function.

Analysis of the Interval ($x \in [2, 3]$)

We now focus our attention on the behaviour of the function within the interval $x = 2$ to $x = 3$. The primary term of interest is $f(x)$, which increases from 0 to 1 over this domain. It is important to note that $g(x)$ remains constant at 1 throughout this interval, implying that the term $2g(x)$ evaluates to 2. This constant contribution, denoted $y_g = 2$, influences the graph of the function in a manner analogous to the role played by $f(x)$ in the previously analysed interval.

Recall that in the preceding interval, $f(x)$ was identically zero, which led to its exclusion from the main functional expression. In contrast, $f(x)$ is now non-zero and increasing, and thus actively contributes to the overall function. Meanwhile, the term $2g(x)$ remains fixed at 2 due to the constancy of $g(x)$.

To better understand the function's behaviour, we construct a table of values based on the expression: $y = 3f^2(x) + 2 + 4 = 3f^2(x) + 6$ which is the current active sub-function.

x	$f(x)$	y
2	0	6
2.25	0.25	6.1875
2.5	0.5	6.75
2.75	0.75	7.6875
3	1	9

Table 5.3.3: Function values over the interval [2, 3]

Calculating the successive differences in the y-values yields the sequence: $0.1875, 0.5625, 0.9375, 1.3125$. The non-uniformity of these differences indicates that the function is not linear but rather exhibits curvature—confirming that it is increasing and nonlinear over this interval.

This observation aligns with the derivative of the function, which was previously determined to be $\frac{dy}{df(x)} = 6f(x)$. For instance, at $f(2.25) = 0.25$, the derivative evaluates to $6 \times 0.25 = 1.5$, and at $f(2.75) = 0.75$, it becomes $6 \times 0.75 = 4.5$. The increasing derivative values further substantiate the function's growth and curvature, thereby validating the correctness of the differentiation.

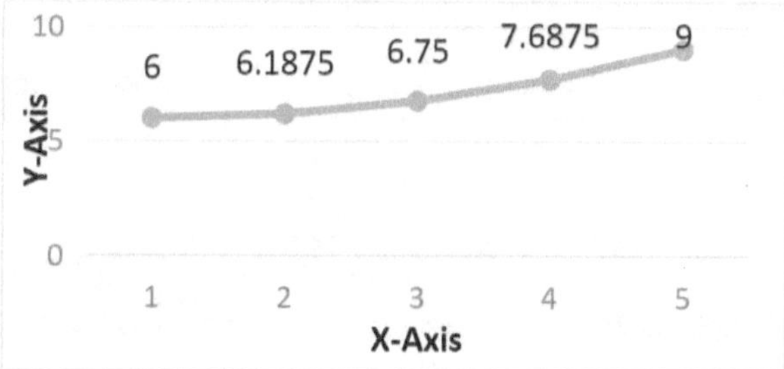

Graph 5.3.2: Curvy segment of the function

As illustrated in **Graph 5.3.1**, the function over this interval is both increasing and curved. When this segment is combined with previous graphical segments, it contributes to the composite structure of the overall function. This piecewise construction is characteristic of what we refer to as a *chain function*—a function composed of multiple interconnected segments, each exhibiting a distinct behaviour.

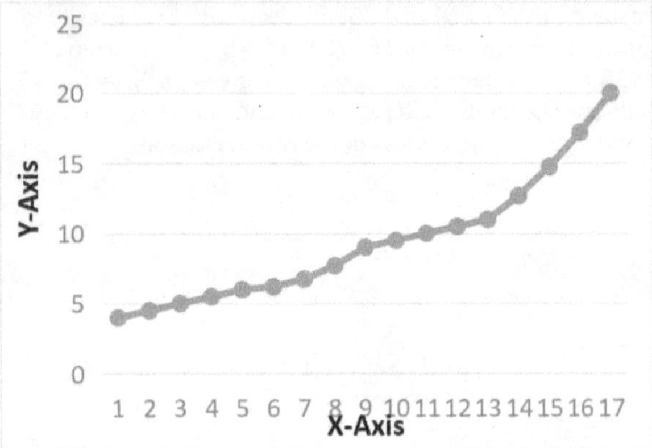

Graph 5.3.3: A visual representation of this composite structure.

Derivatives via the Limit Definition for Chain Functions

We now apply the foundational concept of limits to derive the partial derivatives of a composite function defined in terms of control functions $f(x)$ and $g(x)$. This approach reinforces the correctness of earlier derivative results and deepens our understanding of how these control functions shape the behaviour of the overall function.

Derivative with Respect to $f(x)$

Let the function be:

$$y = 3f^2(x) + 2g(x) + 4$$

We compute the partial derivative with respect to $f(x)$ using the limit definition:

$$\frac{dy}{df(x)} = \lim_{h\to 0}\frac{3(f(x)+h)^2 + 2g(x) + 4 - (3f^2(x) + 2g(x) + 4)}{h}$$

Expanding the numerator:

$$= \lim_{h\to 0}\frac{3\left(f^2(x) + 2f(x)h + h^2\right) - 3f^2(x)}{h}$$

$$= \lim_{h\to 0}\frac{6f(x)h + 3h^2}{h}$$

$$= \lim_{h\to 0}\frac{h(6f(x) + 3h)}{h}$$

$$= \lim_{h\to 0}(6f(x) + 3h)$$

$$= 6f(x)$$

Boxed Summary:

Partial derivative with respect to $f(x)$:

$$\frac{dy}{df(x)} = 6f(x)$$

Derivative with Respect to $g(x)$

Using the same function, we now compute:

$$\frac{dy}{dg(x)} = \lim_{h\to 0}\frac{3f^2(x) + 2(g(x)+h) + 4 - (3f^2(x) + 2g(x) + 4)}{h}$$

$$= \lim_{h\to 0}\frac{2g(x) + 2h - 2g(x)}{h}$$

$$= \lim_{h \to 0} \frac{2h}{h} = 2$$

Boxed Summary:

Partial derivative with respect to $g(x)$:

$$\frac{dy}{dg(x)} = 2$$

Specific Limit Evaluation at $f(x) = 0.75$

Let us verify the derivative at a specific point using the limit definition:

$$\frac{dy}{df(x)} = \frac{dy}{df(2.75)}$$
$$= \lim_{f(x) \to 0.75} \frac{3f^2(x) + 2g(x) + 4 - 7.6875}{f(x) - 0.75}$$

Given $g(x) = 1$, we simplify:

$$= \lim_{f(x) \to 0.75} \frac{3f^2(x) + 2(1) - 3.6875}{f(x) - 0.75}$$

$$= \lim_{f(x) \to 0.75} \frac{3f^2(x) + 1.6875}{f(x) - 0.75}$$

$$= \lim_{f(x) \to 0.75} \frac{(3f(x) + 2.25)(f(x) - 0.75)}{f(x) - 0.75}$$

$$= \lim_{f(x) \to 0.75} (3f(x) + 2,25)$$

$$= 3(0.75) + 2.25 = 4.5$$

This matches the earlier derivative:

$$\frac{dy}{df(x)} = 6f(x) = 6(0.75) = 4.5$$

Conceptual Clarification

Although $f(x)$ and $g(x)$ are functionally dependent on x, they serve as control functions—they determine the values substituted into the main function but are not plotted directly against y. This distinguishes them from independent variables in a 3D function such as $z = 3x^2 + 2y + 4$, which is plotted over (x, y, z). In contrast, our function $y = 3f^2(x) + 2g(x) + 4$ remains in 2D space.

5.4: Exploring the Curvy Function

Let us now examine a new function:

$$y = 2f^2(x) + 3g^2(x) + 5$$

We compute its partial derivatives:

- $\frac{dy}{df(x)} = 4f(x)$
- $\frac{dy}{dg(x)} = 6g(x)$

Both derivatives are positive and vary with x, indicating that the function is composed of increasing curves. The table of values follows:

x	1	2	3	4	5	6	7	8	9
$f(x)$	0	0	1	1	2	2	3	3	4
$g(x)$	0	1	1	2	2	3	3	4	4
y	5	8	10	19	25	40	50	71	85

Table 5.4.1: Values of the curvy function

Final Reflection

Terminology—chain functions or hybrid functions as irregular functions—is apt and original. These functions are composed of distinct segments governed by control functions, and their behaviour is best understood through a combination of limit-based analysis, partial derivatives, and structured tabulation. This framework could form the basis of a new pedagogical approach to modelling piecewise and control-driven systems.

Interval Analysis and Tangent Construction for Control-Driven Functions

We now examine the behaviour of the function across two successive intervals: $\boldsymbol{x} \in [\mathbf{1}, \mathbf{2}]$ and $\boldsymbol{x} \in [\mathbf{2}, \mathbf{3}]$. This analysis reveals how the function transitions between segments and how the control functions $f(x)$ and $g(x)$ influences its curvature and continuity.

Interval $\boldsymbol{x} \in [\mathbf{1}, \mathbf{2}]$: Varying $\boldsymbol{g(x)}$, Constant $\boldsymbol{f(x) = 0}$

In this interval, $f(x)$ remains constant at 0, while $g(x)$ increases from 0 to 1. The function simplifies to: $y = 3g^2(x) + 5$ as a result of these conditions.

x	$g(x)$	y	y'_g
$\mathbf{1}$	0	5	0
$\mathbf{1.2}$	0.2	5.12	1.2
$\mathbf{1.4}$	0.4	5.48	2.4
$\mathbf{1.6}$	0.6	6.08	3.6
$\mathbf{1.8}$	0.8	6.92	4.8
$\mathbf{2}$	1	8	6

Table 5.4.2: Function values and gradients for $x \in [1,2]$

The increasing gradients and non-uniform differences in y-values confirm that the function is curved over this interval.

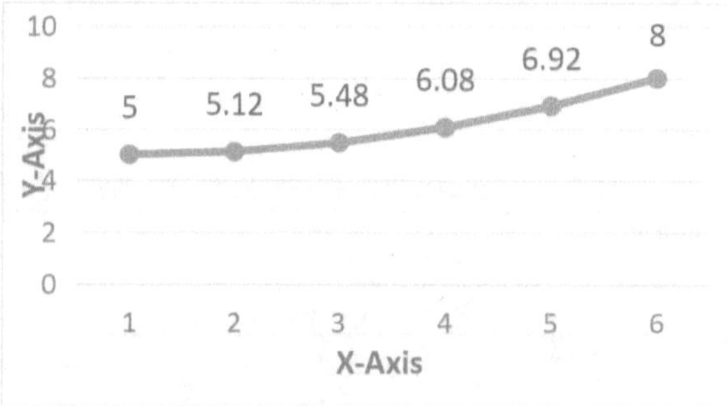

Graph 5.4.1: Part curve at interval $x \in [1, 2]$

Tangent Line at $x = 1.4$

To find the tangent at $x = 1.4$, we use:

- Point: $(g(1.4), y) = (0.4, 5.48)$
- Gradient: $m_g = 6g(x) = 6(0.4) = 2.4$

Using the point-slope form:

$$y - 5.48 = 2.4(g(x) - 0.4)$$

$$y = 2.4g(x) - 0.96 + 5.48$$

$$y = 2.4g(x) + 4.52$$

Boxed Summary:
Tangent at $g(x) = 0.4$:

$$y = 2.4g(x) + 4.52$$

Partial Derivative via Limit Definition

We confirm the derivative with respect to $g(x)$ using the limit:

$$\frac{dy}{dg(x)} = \lim_{h\to 0}\frac{2f^2(x) + 3(g(x)+h)^2 + 5 - (2f^2(x) + 3g^2(x) + 5)}{h}$$

$$= \lim_{h\to 0}\frac{3(g^2(x) + 2g(x)h + h^2) - 3g^2(x)}{h}$$

$$= \lim_{h\to 0}\frac{6g(x)h + 3h^2}{h}$$

$$= \lim_{h\to 0}\frac{h(6g(x) + 3h)}{h}$$

$$= \lim_{h\to 0}(6g(x) + 3h) = 6g(x)$$

Boxed Summary:
Partial derivative with respect to (g(x)):

$$\frac{dy}{dg(x)} = 6g(x)$$

Interval $x \in [2, 3]$: Varying $f(x)$, Constant $g(x)$
In this interval, $g(x)$ is fixed at 1, so the function becomes: $y = 2f^2(x) + 8$.

x	$f(x)$	y	y_f'
2	0	8	0
2.2	0.2	8.08	0.8
2.4	0.4	8.32	1.6
2.6	0.6	8.72	2.4
2.8	0.8	9.28	3.2
3	1	10	4

Table 5.4.3: Function values and gradients for $x \in [2, 3]$

Again, the increasing gradients and curved behaviour confirm that the function is nonlinear and smooth over this interval.

Boxed Definition:
Chain Function (or hybrid function): A function composed of multiple segments, each governed by control functions such as $f(x)$ and $g(x)$, where the overall behaviour is determined by the piecewise interaction of these controls. These functions often exhibit curvature, discontinuities, or non-standard transitions, making them ideal for modelling dynamic systems with localized behaviour.

Final Partial Limit and Composite Graph of the Curvy Chain Function

We now compute the partial limit of the function with respect to $f(x)$ for the expression:

$$2f^2(x) + 3g^2(x) + 5$$

Partial Derivative via Limit Definition

Using the limit definition:

$$\frac{dy}{df(x)}$$
$$= \lim_{h \to 0} \frac{2(f(x)+h)^2 + 3g^2(x) + 5 - (2f^2(x) + 3g^2(x) + 5)}{h}$$

Expanding:

$$= \lim_{h \to 0} \frac{2(f^2(x) + 2f(x)h + h^2) - 2f^2(x)}{h}$$

$$= \lim_{h \to 0} \frac{4f(x)h + 2h^2}{h}$$

$$= \lim_{h \to 0} \frac{h(4f(x) + 2h)}{h}$$

$$= \lim_{h \to 0} (4f(x) + 2h) = 4f(x)$$

Boxed Summary:
Partial derivative with respect to $f(x)$:

$$\frac{dy}{df(x)} = 4f(x)$$

This confirms the earlier derivative obtained via the power rule and reinforces the curvature of the function over the interval.

Composite Graphical Interpretation

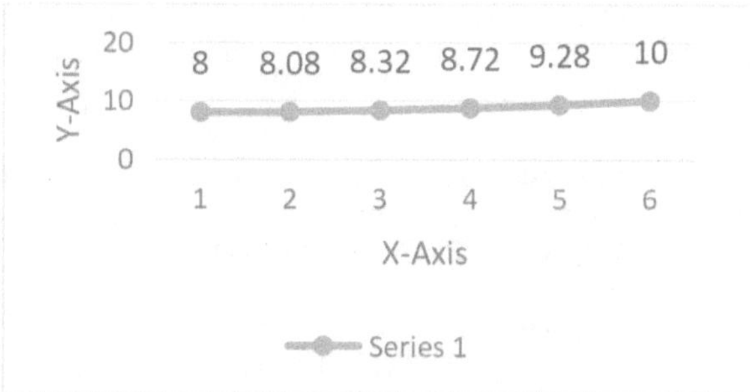

Graph 5.4.2: Part curve at the internal $x \in [2, 3]$

The graph illustrates the behaviour of the function over the interval where $f(x)$ varies and $g(x)$ remains constant. The curvature is evident from the increasing derivative values and the non-linear shape of the graph.

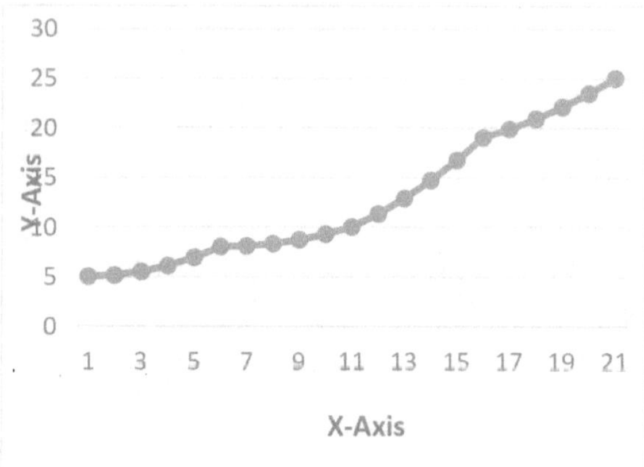

Graph 5.4.3: bringing together the pieces of the function at $x \in [1, 5]$

The graph presents the full composite structure of the function over the domain $x \in [1, 5]$. This graph combines four distinct segments, each governed by different behaviors of the control functions $f(x)$ and $g(x)$. These segments are:

1. **Segment 1**: $f(x) = 0$, $g(x)$ increasing → governed by $y = 3g^2(x) + 5$
2. **Segment 2**: $g(x) = 1$, $f(x)$ increasing → governed by $y = 2f^2(x) + 8$
3. **Segment 3**: Both $f(x)$ and $g(x)$ increasing → governed by $2f^2(x) + 3g^2(x) + 5$
4. **Segment 4**: Further extension of the control functions → governed by a continuation of the composite rule

Each segment contributes a unique curvature and gradient profile, forming a continuous but non-uniform structure—what we now formally recognize as a **chain function**.

Final Reflection: The Nature of Chain Functions

The work has illuminated a class of functions that blend piecewise behaviour with control-driven dynamics. These functions are not merely irregular—they are *structured composites*, where each segment is governed by a distinct rule derived from control functions.

Boxed Definition (Refined):
Chain Function: A function composed of multiple segments, each defined by control functions such as $f(x)$ and $g(x)$, which determine the local behavior and curvature of the function. These segments are joined continuously or piecewise, forming a composite structure that models dynamic systems with localized control.

Chapter 6

6.1: Transformations of Irregular Functions

We now examine transformations applied to an irregular function of the form $y = f(x)$, where $f(x)$ is a control function. Unlike standard algebraic functions, control functions often exhibit piecewise or non-uniform behaviour, making them ideal for modelling real-world processes. The general transformation framework we adopt is:

$$y = af(bx \pm c) \pm d$$

This form allows us to explore how the parameters a, b, c, and d affect the graph of $f(x)$. To illustrate these effects, we consider five specific transformations:

- $y = f(x)$ — the base function
- $y_1 = 3f(x)$ — vertical stretch
- $y_2 = f(x + 1)$ — horizontal shift
- $y_3 = f(3x)$ — horizontal compression
- $y_4 = f(x) - 2$ — vertical shift downward

We assume a cardinality of x to be 2, meaning each value of $f(x)$ repeats twice across successive x-values.

Table 6.1.1: Base Function $y = f(x)$

x	1	2	3	4	5	6	7
$f(x)$	1	1	2	2	3	3	4
$y = f(x)$	1	1	2	2	3	3	4

This table defines the control function and serves as the reference for all subsequent transformations.

Table 6.1.2: Vertical Stretch $y_1 = 3f(x)$

x	1	2	3	4	5	6	7
$f(x)$	1	1	2	2	3	3	4
y_1	3	3	6	6	9	9	12

The coefficient $a = 3$ scales the output vertically.

This transformation multiplies each output value by 3, stretching the graph vertically.

Table 6.1.3: Horizontal shift $y_2 = f(x + 1)$

x	1	2	3	4	5	6	7
$f(x)$	1	1	2	2	3	3	4
$f(x + 1)$	$f(2)$	$f(3)$	$f(4)$	$f(5)$	$f(6)$	$f(7)$	$f(8)$
y_2	1	2	2	3	3	4	4

The term $c = 1$ induces a leftward shift.

This shifts the graph of $f(x)$ one unit to the left.

The graph below shows the above functions drawn in the same system of axes.

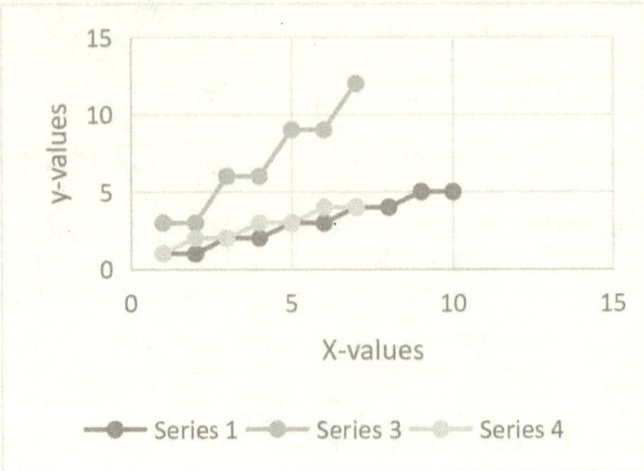

Graph 6.1.1: The graph of the above three functions drawn

Table 6.1.4: Horizontal compression $y_3=f(3x)$

x	**1**	**2**	**3**	**4**	**5**	**6**	**7**
$f(x)$	1	1	2	2	3	3	4
$f(3x)$	$f(3)$	$f(6)$	$f(9)$	$f(12)$	$f(15)$	$f(18)$	$f(21)$
y_3	2	3	5	6	8	9	11

The coefficient $b = 3$ expands the graph horizontally.

This transformation accelerates the input progression, resulting in a compressed graph along the horizontal axis.

Table 6.1.5: Vertical shift $y_4=f(x)$-2

x	**1**	**2**	**3**	**4**	**5**	**6**	**7**
$f(x)$	1	1	2	2	3	3	4
y_4	-1	-1	0	0	1	1	2

The constant $d = -2$ shifts the graph downward: This transformation reduces each output value by 2 units.

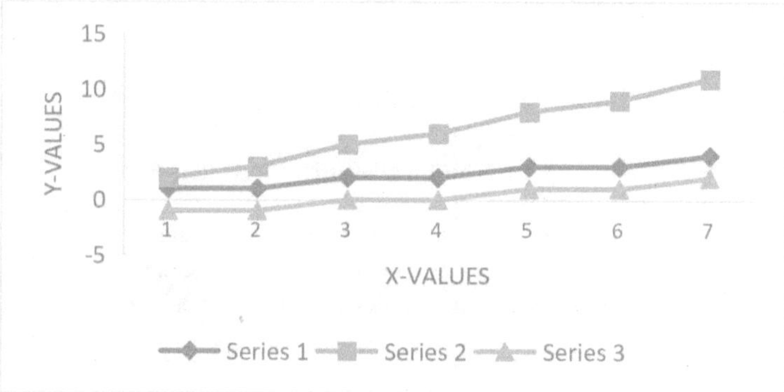

Graph 6.1.2: Graphs of the last two functions and the main function y=f(x)

Interpretation of Transformations

- **Vertical Stretch (a)**: Amplifies the output values, increasing the graph's height. The graph is the top one on graph 6.1.1 (series 3).
- **Horizontal Shift (c)**: Alters the input domain, translating the graph left or right. The graph is middle one on top of the base graph on graph 6.1.1 (i.e. series 4).
- **Horizontal Compression (b)**: Modifies the input scale, compressing or expanding the graph horizontally. Graph on top of base graph on graph 6.1.2 (series 2).
- **Vertical Shift (d)**: Adjusts the output baseline, raising or lowering the graph. The graph for this is below base function graph on graph 6.1.2 (series 3)

These transformations illustrate how control functions—though irregular—respond predictably to algebraic manipulation, enabling flexible modeling of real-world processes.

6.2. Other Transformations

6.2.1. The Two-Gradient Irregular Function

We further initiate our exploration of irregular transformations by examining a linear function composed of two control functions with equal gradients, specifically:

$$y = f(x) + g(x)$$

In this formulation, both $f(x)$ and $g(x)$ possess gradients of one, and their additive combination yields a function governed by dual control mechanisms. For contrast, we also consider the analogous function with a subtractive relationship:

$$y = f(x) - g(x)$$

Throughout this analysis, we maintain a fixed cardinality of the domain, $|x| = 2$, and impose the alignment inequality:

$$g(x) \leq f(x) < x$$

These constructions closely parallel the methods presented by Magnus (2020) in her video titled *Transforming Irregular Functions*. To facilitate visualization, we begin with tabular representations of both functions.

Table 6.2.1.1: Values for $y = f(x) + g(x)$

x	1	2	3	4	5	6	7	8	9
$f(x)$	0	1	1	2	2	3	3	4	4
$g(x)$	0	0	1	1	2	2	3	3	4
y	0	1	2	3	4	5	6	7	8

Table 6.2.1.2: Values for (y = f(x) - g(x))

x	1	2	3	4	5	6	7	8	9
$f(x)$	0	1	1	2	2	3	3	4	4
$g(x)$	0	0	1	1	2	2	3	3	4
y	0	1	0	1	0	1	0	1	0

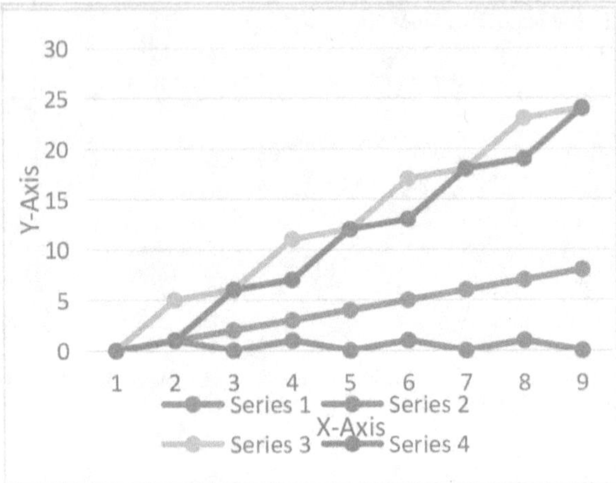

Graph 6.2.1.1: Illustrates both functions on a shared coordinate system.

The graph of $y = f(x) + g(x)$ (Series 1, second from the bottom, an increasing straight line) exhibits linear behaviour with an intercept at $x = 1$ and a constant gradient of one, resembling the standard line $y = x - 1$. In contrast, the graph of $y = f(x) - g(x)$ (Series 2, the first at the bottom, exhibit joint lines that go up and down) oscillates between 0 and 1, reflecting the alternating influence of the two gradients (of 1 and -1).

To examine the impact of scalar multiplication, we also plot the functions:

- $y = 5f(x) + g(x)$ (Series 3, the fourth graph from the bottom or first from the top (increasing and zigzag in shape))
- $y = f(x) + 5g(x)$ (Series 4, third graph from the bottom or second from the top (increasing and zigzag in shape))

These functions demonstrate increased steepness due to the introduction of a gradient of 5. The resulting graphs consist of piecewise linear segments alternating between gradients of 5 and 1. Specifically, in $y = 5f(x) + g(x)$, the initial segment adopts gradient 5, followed by gradient 1, and so forth. The function $y = f(x) + 5g(x)$ exhibits the same alternation but begins with gradient 1.

This alternating gradient structure leads to a composite function formed by successive linear segments. If the gradients are denoted d_1 and d_2, the sequence of segments follows the pattern: $(d_1, d_2, d_1, d_2, \ldots)$. The fluctuation observed in $y = f(x) - g(x)$ arises from the interplay between gradients of 1 and -1, effectively fragmenting a single line into alternating directional components.

Such behaviour is frequently encountered in applications involving binary switching mechanisms, where the function alternates between states—commonly represented by 0 and 1. This is particularly relevant in fields such as computer science and electronics, where control functions model digital signals and switching behaviour.

The graphical behaviour of these functions is determined by the signs and magnitudes of their constituent gradients. Increasing the number of distinct gradients introduces additional segments, each contributing to the overall shape and complexity of the function.

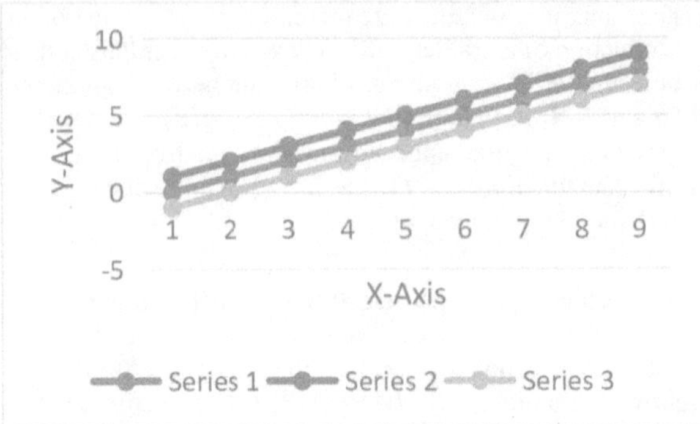

Graph 6.2.1.2: This graph extends this analysis by incorporating constant shifts:

Identifying graphs in **Graph 6.2.1.2**

- $y = f(x) + g(x)$ (Series 1, middle graph)
- $y = f(x) + g(x) + 1$ (Series 2, top graph)
- $y = f(x) + g(x) - 1$ (Series 3, bottom graph in)

The addition of a constant translates the graph vertically. Specifically, adding one shifts the graph upward by one unit, while subtracting one shifts it downward. These transformations preserve the gradient structure while modifying the function's range.

6.2.2. Addition-Based Transformations of Sigma Functions

We now turn our attention to summation-based functions, and those involving nested summations. For convenience, we refer to these as *sigma functions*, owing to their use of the summation symbol $\sum$ and their intrinsic role in aggregating values. It is important to note that the sigma functions discussed here differ from conventional summation functions encountered elsewhere in mathematics although similar to partial sums.

The gradient equation or function can be irregular as in:

$$y = -1 + \sum_{i=1}^{x-1} \frac{(-1)^{i-1}(4f(i) + g(i) + 5)}{2}$$

Its associated gradient sequence is given by:

$$d_i = \frac{(-1)^{i-1}(4f(i) + g(i) + 5)}{2}$$

This gradient is both alternating and irregular, exemplifying the non-standard behaviour of such sigma functions.

Let us now examine the transformation properties of the function:

$$y = \sum_{i=1}^{x-1} (-1)^i (i + 1)$$

This sigma function exhibits a regular alternating gradient. The summation itself serves as a control function, as illustrated in Table 6.2.2.1. Although the gradient alternates, it remains regular when considering only its positive terms.

Table 6.2.2.1: Values of the Sigma Function

x	1	2	3	4	5	6	7	8	9
S_x	0	-2	1	-3	2	-4	3	-5	4
y	0	-2	1	-3	2	-4	3	-5	4

Now consider the effect of adding a constant to the sigma function. If we add 1:

$$y = 1 + \sum_{i=1}^{x-1} (-1)^i (i+1)$$

We obtain the values shown in Table 6.2.2.2:

Table 6.2.2.2: Sigma Function Shifted Upward

x	1	2	3	4	5	6	7	8	9
S_x	0	-2	1	-3	2	-4	3	-5	4
y	1	-1	2	-2	3	-3	4	-4	5

Subtracting the sigma function from 1 yields:

$$y = 1 - \sum_{i=1}^{x-1} (-1)^i (i+1)$$

Resulting in Table 6.2.2.3:

Table 6.2.2.3: Sigma Function Inverted

x	1	2	3	4	5	6	7	8	9
S_x	0	-2	1	-3	2	-4	3	-5	4
y	1	3	0	4	-1	5	-2	6	-3

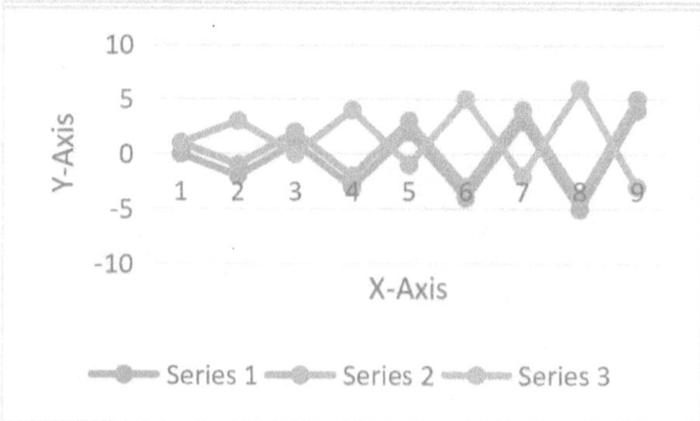

Graph 6.2.2.1: This graph illustrates these three transformations.

The original sigma function (Series 1, the graph below on Graph 6.2.2.1) is shifted upward by 1 unit when added to 1 (Series 2, second graph above Series 1), and inverted when subtracted from 1 (Series 3, top graph that's inverted to the other first two).

Similarly, adding -1 to the sigma function:

$$y =- 1 + \sum_{i=1}^{x-1} (-1)^i (i + 1)$$

Produces the values in Table 6.2.2.4:

Table 6.2.2.4: Sigma Function Shifted Downward

x	1	2	3	4	5	6	7	8	9
S_x	0	-2	1	-3	2	-4	3	-5	4
y	-1	-3	0	-4	1	-5	2	-6	3

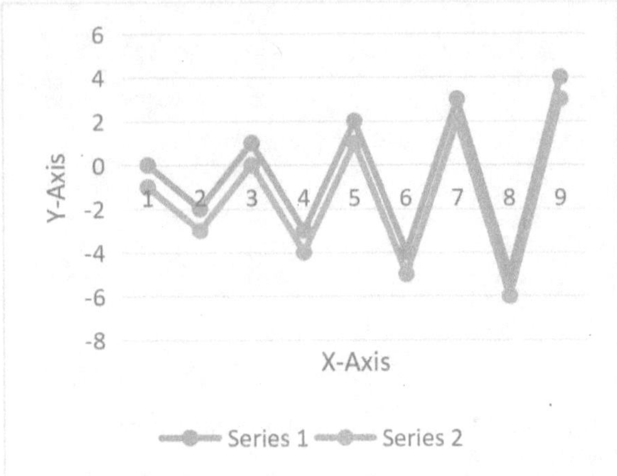

Graph 6.2.2.2

The graph confirms that the sigma function is translated downward by one unit. Thus, adding or subtracting a constant result in vertical shifts, with the sign of the constant determining the direction.

The gradient of the sigma function is:

$$\frac{dy}{dx} = (-1)^i(i + 1)$$

Let us now explore how modifying this gradient affects the sigma function. Rewriting the original function for reference:

$$y = \sum_{i=1}^{x-1} (-1)^i(i + 1)$$

Changing the summand to $i - 1$ yields:

$$y = \sum_{i=1}^{x-1} (-1)^i (i - 1)$$

Table 6.2.2.5: Transformation via Subtraction in Summand

x	1	2	3	4	5	6	7	8	9
S_x	0	0	1	-1	2	-2	3	-3	4
y	0	0	1	-1	2	-2	3	-3	4

Next, we multiply the index i by 3 (in the summand):

$$y = \sum_{i=1}^{x-1} (-1)^i (3i + 1)$$

Table 6.2.2.6: Transformation via Multiplication by 3

x	1	2	3	4	5	6	7	8	9
S_x	0	-4	3	-7	6	-10	9	-13	12
y	0	-4	3	-7	6	-10	9	-13	12

Multiplying by -3 instead:

$$y = \sum_{i=1}^{x-1} (-1)^i (- 3i + 1)$$

Table 6.2.2.7: Transformation via Multiplication by -3

x	1	2	3	4	5	6	7	8	9
S_x	0	2	-3	5	-6	8	-9	11	-12
y	0	2	-3	5	-6	8	-9	11	-12

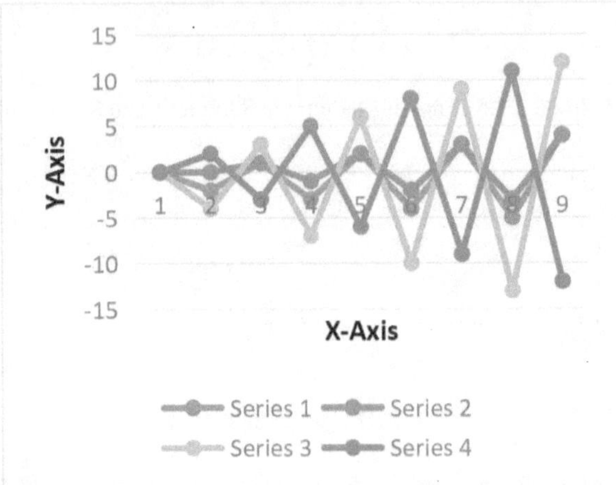

Graph 6.2.2.3

The graph displays the original sigma function (Series 1, second graph from the bottom (you can see it on even x values)) alongside its transformations. The third graph from the bottom (Series 2, seen on even x values) reflects a vertical compression of the negative values, shifting them closer to zero. The first graph from the bottom (Series 3, seen on even x values), resulting from multiplication by 3, shows vertical stretching without inversion. In contrast, multiplication by -3 yields both stretching and inversion.

These sigma functions and their transformations are examined here due to their distinct behaviour compared to traditional series encountered in calculus. Notably, the use of partial sums—dependent on the discrete variable x—is central to their structure and interpretation.

6.2.3. Double Sigma Addition Functions

This section examines a transformation involving a function defined by a double summation:

$$y = \sum_{i=1}^{x-1} (1 + \sum_{j=1}^{i-1} (-1)^j (j + 1))$$

This expression may alternatively be represented as:

$$y = \sum_{i=1}^{x-1} 1 + \sum_{i=1}^{x-1} \sum_{j=1}^{i-1} (-1)^j (j + 1)$$

The initial formulation can be interpreted as a factorized version of the latter, where the outer summation symbol has been distributed across both terms. This structural equivalence invites further exploration of how sign alterations affect the behaviour of the function.

To investigate this, we consider two transformations:

1. Replacing the constant 1 with -1
2. Changing the sign following the constant 1 from addition to subtraction

We begin by analysing the original function through tabulated values:

Table 6.2.3.1: Original Function

x	1	2	3	4	5	6	7	8	9
S_x	0	1	0	2	0	3	0	4	0
S_i	0	0	-2	1	-3	2	-4	3	-5
y	0	1	0	2	0	3	0	4	0

Here, S_x denotes the external summation (i.e., the full expression), while S_i captures the internal summation component. The absence of an initial value when the sum is zero reflects the non-contributory nature of the base case.

We now examine the transformation:

$$y = \sum_{i=1}^{x-1} (1 - \sum_{j=1}^{i-1} (-1)^j (j+1))$$

Table 6.2.3.2: Subtractive Internal Sum

x	1	2	3	4	5	6	7	8	9
S_x	0	1	4	4	8	7	12	10	16
S_i	0	0	-2	-1	-4	-2	-6	-3	-8
y	0	1	4	4	8	7	12	10	16

Next, we consider the transformation where the constant is replaced with -1:

$$y = \sum_{i=1}^{x-1} \left(-1 + \sum_{j=1}^{i-1} (-1)^j (j+1) \right)$$

Table 6.2.3.3: Negative Constant with Additive Internal Sum

x	1	2	3	4	5	6	7	8	9
S_x	0	-1	-4	-4	-8	-7	-12	-10	-16
S_i	0	0	-2	-1	-4	-2	-6	-3	-8
y	0	-1	-4	-4	-8	-7	-12	-10	-16

The results reveal a striking symmetry: the transformed functions in Tables 6.2.3.2 and 6.2.3.3 are sign inverses of one another. When one yields positive values, the other produces their negative counterparts.

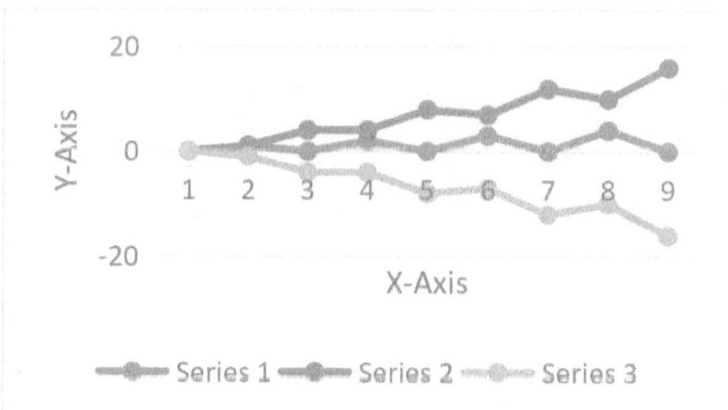

Graph 6.2.3.1: Graphical Representation of the Transformation Discussed so far

The graphical representations further reinforce this duality. The top curve (Series2) exhibits a steep positive incline, while the bottom curve (Series3) mirrors this behaviour with a steep negative slope. These two series are symmetric about the x-axis (i.e., the line $y = 0$). The middle curve (Series1), corresponding to the original summation function, displays a rising pattern and serves as the baseline for comparison.

Boxed Summary: Key Observations

- The double sigma function exhibits alternating behaviour depending on sign placement.
- Graphical symmetry emerges when constants are negated or internal sums are subtracted.
- The original function acts as a neutral baseline, with transformed variants diverging symmetrically.

Chapter 7

7.1 Application Examples in Commerce

This chapter explores practical applications of some of the irregular sequences introduced in Chapter 1, with a particular focus on how such sequences and their associated functions can model real-world scenarios—especially those involving operational constraints or inefficiencies. The examples presented here demonstrate the utility of irregular sequences in contexts where conventional models fall short.

7.1.1. Application example: Cake Packing and Cost Modelling

Consider a packaging scenario involving two types of cakes: chocolate and ginger. The chocolate cake is priced at R4, while the ginger cake costs R1. Each cake requires one minute to pack, and due to human or mechanical limitations, the packer cannot simultaneously pack both cakes. That is, while packing a chocolate cake, the ginger cake must remain unpacked, and vice versa. The cakes are packed in packets where each packet contains the two different cakes.

Suppose the packing process begins with a balance of R5. The total cost of packed cakes over time can be modelled by the function:

$$y = 4f(x) + g(x) + 5$$

Here:

- $f(x)$ denotes the number of chocolate cakes packed by time x,
- $g(x)$ denotes the number of ginger cakes packed by time x,

- The coefficients 4 and 1 represent the respective prices of the cakes,
- The constant 5 is the initial balance.

Table 7.1.1.1: Packing Progress Over Time

$time(x)$ **in min**	$number\ of$ **choc cakes** $f(x)$	$number\ of$ **ginger cakes** $g(x$
1	0	0
2	1	0
3	1	1
4	2	1
5	2	2

This table illustrates the sequential nature of the packing process. At $x = 1$, no cakes have been packed, and the total cost is simply the initial balance: $y = 5$. At $x = 2$, one chocolate cake is packed, increasing the cost to $y = 9$. At $x = 3$, a ginger cake is added, resulting in $y = 10$. The packing alternates due to the constraint that only one cake can be packed at a time and the cakes are packed interchangeably.

The structure of the cost function follows the general form:

$$y = d_1 f(x) \pm d_2 g(x) \pm a$$

Where:

- y is the total cost of packed cakes,
- $f(x)$ and $g(x)$ are control functions representing the number of cakes packed,
- d_1 and d_2 are the respective unit costs (R4 and R1),
- a is the initial balance (R5).

Why Use Control Functions?

Instead of representing the number of cakes with a single variable, we employ control functions $f(x)$ and $g(x)$ to reflect the operational constraint: only one cake can be packed at a time. These functions allow us to model the packing process with temporal granularity and account for delays or suspensions in activity. In this context, $f(x)$ and $g(x)$ are not merely counting functions—they are *control functions*, regulating the input values based on task feasibility.

This modelling approach is particularly useful in scenarios involving limited multitasking capacity. Even if two individuals are involved, the packing of both cakes into a single packet still occurs sequentially. The control functions enable us to blend these tasks into a unified cost function that reflects the cumulative output.

Table 7.1.1.2: Extended Packing and Cost Analysis

x	1	2	3	4	5	6	7	8
$f(x)$	0	1	1	2	2	3	3	4
$g(x)$	0	0	1	1	2	2	3	3
y	5	9	10	14	15	19	20	24

From this table, we can answer the earlier question: *What is the total cost after 5 minutes of packing?* The value from the table is $y = 15$, which includes the initial balance of R5. This implies that two complete packets have been packed, each costing R5 (R4 for the chocolate cake and R1 for the ginger cake), totalling R10, plus the initial R5.

Derivatives and Unit Costs

To extract the unit costs from the cost function, we apply partial differentiation:

Given:

$$y = 4f(x) + g(x) + 5$$

- The partial derivative with respect to $f(x)$ yields: $\frac{\partial y}{\partial f(x)} = 4$. This confirms the unit cost of the chocolate cake.
- The partial derivative with respect to $g(x)$ yields: $\frac{\partial y}{\partial g(x)} = 1$. This confirms the unit cost of the ginger cake.

Summary

This example illustrates how irregular sequences and control functions can be applied to model constrained, sequential tasks in commercial settings. By embedding operational limitations into the structure of the function, we obtain a more realistic and adaptable representation of cost and productivity.

7.1.2. Modelling the Coffee-Making Process Using the Equation $\mathbf{y = df(x) = 2f(x)}$

To explore the dynamics of making tea (or coffee) in a café setting, we consider the equation $y = 2f(x)$, where the constant $d = 2$ represents the price per cup of coffee, inclusive of profit and VAT. This constant serves as the rate of return per unit of output. We begin with an initial account balance of zero.

Let us examine the following table:

x	1	2	3	4	5	6	7
$f(x)$	0	1	1	2	2	3	3
y	0	2	2	4	4	6	6

Table 7.1.2.1: Output and earnings based on time and production

Here:

- x denotes the time (in minutes) spent making coffee.
- $f(x)$ is a function that maps time to the number of cups produced.
- $y = 2f(x)$ represents the total monetary value of the coffee made.

The function (f) encapsulates the job of coffee-making: it receives time as input and returns the number of cups produced. The total earnings y are then computed by multiplying the number of cups by the unit price $d = 2$.

Comparison with the Linear Model $y = 2x$

To assess the suitability of $y = 2f(x)$, we compare it with the simpler linear model $y = 2x$, which assumes a direct proportionality between time and earnings.

x	1	2	3	4	5	6	7
y	2	4	6	8	10	12	14

Table 7.1.2.2: Linear earnings model assuming constant productivity

Analysis of $y = 2f(x)$

In the first minute, no coffee is produced, hence no earnings. By the second minute, one cup is completed, yielding R2. The third minute reflects a continuation of the previous output, allowing time for delivery or auxiliary tasks. This model accommodates the realistic pacing of a single worker who alternates between production and service.

The function $f(x)$ captures the variability in productivity over time, acknowledging that not every minute yields a new cup. This is particularly relevant in a setting where one individual is responsible for both preparation and delivery, making the model time-consistent for a single worker.

Limitations of $y = 2x$

The linear model $y = 2x$ assumes that each minute results in a new cup, which oversimplifies the process. It fails to account for delivery time, worker variability, and task switching. To model these aspects accurately, one would need multiple equations:

- One linking time to cups produced
- Another linking cups to revenue
- Possibly a third to allocate time between preparation and delivery

This complexity makes $y = 2x$ less practical for modelling individual performance in a realistic café environment.

Conclusion

The equation $y = 2f(x)$ offers a more nuanced and integrated representation of the coffee-making process. It accommodates fluctuations in productivity, delivery time, and individual pacing. Unlike ($y = 2x$), which generalizes across multiple workers and assumes uniform output, ($y = 2f(x)$) enables tracking of individual performance and aligns more closely with operational realities. By embedding both time and output into a single function, it simplifies the modelling of total earnings and enhances interpretability.

Comparing Single-Variable and Multi-Variable Function Models in Cake Packing

We consider two candidate models for representing the total quantity of cakes packed in a bakery setting:

- **Model A**: $z = 4x + y + 5$, a multivariable function mapping inputs x and y to an output z.
- **Model B**: $y = 4f(x) + g(x) + 5$, a single-variable composition where both f and g are functions of time x.

Structural Differences

Model A involves three variables—x, y, and z—and treats x and y as independent inputs. In contrast, Model B operates with two variables, where x is the independent variable and y is the dependent output derived from the combined behaviour of $f(x)$ and $g(x)$.

This distinction is critical: Model B reflects a time-dependent process where both cake types (e.g., chocolate and ginger) are packed sequentially or conditionally based on time, whereas Model A assumes simultaneous and independent contributions from x and y, which may not align with the operational constraints of the packing task.

Practical Implications

Attempting to apply Model A (($z = 4x + y + 5$) introduces ambiguity in job allocation. Specifically, it lacks a mechanism to encode **cardinality constraints**—such as the rule that chocolate cakes and ginger cakes are not packed concurrently. This absence of temporal sequencing or conditional logic leads to misrepresentation of the task structure.

Model B, by contrast, allows for a more nuanced representation. The functions $f(x)$ and $g(x)$ can be defined to reflect distinct packing phases or conditional outputs, such as:

- $f(x)$: number of chocolate cakes packed at time x
- $g(x)$: number of ginger cakes packed at time x, possibly zero if chocolate cakes are being packed

This approach respects the job definition and sequencing constraints, enabling accurate modelling of individual worker behaviour and task allocation over time.

Conclusion

While multivariate functions like $z = 4x + y + 5$ offer generality, they may obscure the temporal and conditional structure inherent in real-world tasks such as cake packing. The single-variable composite model $y = 4f(x) + g(x) + 5$ is better suited to capturing the dynamics of sequential production, especially when specific rules govern task execution. It provides a more faithful representation of the process and supports clearer allocation of responsibilities.

Generalizing to Other Packing Problems

1. Sequential or Conditional Packing Tasks

In many real-world packing problems, certain items must be packed before others, or cannot be packed simultaneously (as with your chocolate vs. ginger cake example). A function-based model like:

$$y = af(x) + bg(x) + c$$

Allows you to:

- Assign distinct functions to different item types or stages
- Encode conditional logic (e.g., $g(x) = 0$ when $f(x) > 0$)
- Track cumulative output over time

This is especially useful in assembly lines, food prep, or logistics where sequencing matters.

2. Single Worker vs. Multi-Worker Dynamics

Our emphasis on time-consistency for a single worker is key. Function-based models preserve individual pacing and allow for personalized performance tracking. In contrast, multivariable models like $z = 4x + y + 5$ often assume parallel processing, which may obscure individual contributions.

For example:

- $f(x)$: number of boxes packed by Worker A at time x
- $g(x)$: number of boxes packed by Worker B at time x

You can then compare or combine outputs while maintaining temporal integrity.

3. Resource Allocation and Switching Costs

Packing problems often involve switching between tasks or tools (e.g., changing packaging materials, reconfiguring machinery). Function-based models can incorporate switching delays or penalties:

$$y = 4f(x) + g(x - \Delta) + 5$$

Here, Δ represents a delay or transition time between tasks — a powerful way to model inefficiencies or setup costs.

4. Inventory and Demand Constraints

Suppose you have limited stock or fluctuating demand. You can define:

- $f(x)$: number of items packed based on available inventory
- $g(x)$: number of items packed based on incoming orders

Then use a composite function to model fulfilment:

$$y = \min (f(x), g(x))$$

This captures real-world constraints elegantly.

Example: Packing Electronics

Imagine a warehouse packing smartphones and tablets. You could define:

- $f(x)$: smartphones packed per hour
- $g(x)$: tablets packed per hour
- $y = 300f(x) + 500g(x) + 100$: total value of packed goods (in Rands)

This model lets you:

- Track earnings
- Optimize packing schedules
- Allocate workers based on item value or demand

Other applications with similarities

1. Superposition Principle (Linear Systems Theory)

- Core Idea: In linear systems, the total output is the sum of individual outputs caused by each input.

- Connection: Our model treats each packing task (e.g., chocolate vs. ginger cake) as a separate input function. The total packed output is a linear combination—exactly what superposition describes.

- Used In: Signal processing, electrical circuits, differential equations.

2. Time-Dependent Control Functions (Control Theory)

- Core Idea: Control functions govern system behaviour over time, often with constraints or switching logic.

- Connection: Our use of f(x) and g(x) as time-indexed packing rates mirrors control inputs in dynamic systems. Especially relevant when modelling task switching, delays, or prioritization.

- Used In: Robotics, manufacturing systems, optimization.

7.1.3. Cost Modelling in Factory Production Using Control Functions

In this section, we examine a cost model for item production within a factory setting. The objective is to determine the total monetary value, in Rand(s), of items produced over time, using a two-variable functional equation:

$$y = [(x-1) - f(x)]d_1 + g(x)d_2 + a$$

By substituting the constants $d_1 = 200$, $d_2 = 100$, and $a = 500$, we obtain the working equation:

$$y = [(x-1) - f(x)]200 + 100g(x) + 500$$

Variable Definitions

- x: Time in hours required to produce a single item
- $f(x)$: Quantity of the first item (hereafter referred to as the *f-item*)
- $g(x)$: Quantity of the second item (hereafter referred to as the *g-item*)
- $d_1 = 200$: Unit cost of producing an *f-item*
- $d_2 = 100$): Unit cost of producing a *g-item*
- $a = 500$: Initial balance or base cost
- y: Total production cost in Rand(s)
- The cardinality of x is 4 and $x \geq f(x) = g(x)$

The term $200(x-1)$ captures the cumulative increase in cost due to time progression, while $200f(x)$ accounts for the repeated production of *f-items*. The interplay between these terms reflects both temporal and quantitative dynamics in the production process.

Table 7.1.3.1: Sample Production Data

x in hours	1	2	3	4	5	6	7	8	9
$f(x)$	1	1	1	1	2	2	2	2	3
$g(x)$	1	1	1	1	2	2	2	2	3
y in Rand(s)	400	600	800	1000	1100	1300	1500	1700	1800

Interpretation of the Term $[(x-1)-f(x)]200$

This component represents a dynamic adjustment in the cost associated with *f-items*. A negative value implies either:

1. The removal of an *f-item* for testing purposes, or
2. A deficit in production, necessitating a deduction from the current balance.

This term thus serves as a control mechanism, regulating the inclusion or exclusion of *f-items* in the cost computation.

Production Pattern

The model exhibits a cyclical pattern:

- Every three additions of *f-items* are followed by a single addition of a *g-item*.
- Conversely, each *g-item* addition is followed by three *f-item* additions.

This alternating structure ensures a balanced progression in production, with cost adjustments reflecting the timing and quantity of each item type.

Example 4.1: Cost Calculation at $x = 9$ Hours

Given:

- Cardinality of x: 4
- Alignment inequality: $x \geq f(x) = g(x)$

Step 1: Derive the control function for $f(x)$
Assuming a non-recurrent natural sequence, we use:

$$x = 4f(x) - 3$$

Where

$$f(x) = \frac{x+3}{4}$$

Substituting $x = 9$:

$$f(x) = \frac{9+3}{4} = 3$$

From the alignment inequality, $g(x) = f(x) = 3$

Step 2: Substitute into the main equation:

$$y = [(9-1) - 3]200 + 100(3) + 500$$

$$y = 5(200) + 300 + 500$$

$$y = 1800$$

Thus, the total production cost after 9 hours is **R1800**.

Generalization for Extended Time Frames

When $f(x)$ and $g(x)$ are unknown, auxiliary information—such as cardinality and alignment inequalities—can be leveraged to derive control functions. A general form relating x and $f(x)$ may be expressed as:

$$x = |x| \times f(x) \pm c$$

where c is a constant determined by the production structure.

7.1.4: Modelling Business Profit and Investment Using a Summation Function

In this section, we examine a commercial scenario involving alternating cycles of stock investment and profit realization. The model employs a summation function to track the evolution of a business account balance over discrete time intervals, measured in months.

Main Equation

We define the total account balance y as:

$$y = 900 + \sum_{i=1}^{x-1} (-1)^i (600f(i) + 100g(i) + 700)$$

This alternating summation reflects the periodic inflow and outflow of funds, where:

- Positive terms represent revenue from stock sales and profit realization.
- Negative terms represent expenditures on stock purchases and capital injections.

Variable Definitions

Symbol	Description
x	Time in months
i	Index of summation (month counter)
$f(i)$	Multiples of profit-generating transactions
$g(i)$	Multiples of capital investment transactions
d_1	Net profit rate associated with $f(i)$
d_2	Capital increase rate associated with $g(i)$
y	Business account balance in Rands

Table 7.1.4.1: Monthly Progression

x	1	2	3	4	5	6	7	8	9
$f(x)$	0	0	1	1	2	2	3	3	4
$g(x)$	0	0	0	1	1	2	2	3	3
$\sum$	0	-7	6	-8	12	-9	18	-10	24
y	9	2	15	1	21	0	27	-1	33

All the amounts (for $\sum$ and y) end in double zeros (00), e.g. -7 is -700 or 15 is 1500.

Interpretation of the Model

At $x = 1$, the business begins with an initial balance of R900. No transactions occur, so both $f(i)$ and $g(i)$ are zero. At $x = 2$, the business purchases stock worth R700, reducing the balance to R200. By $x = 3$, the stock is sold, generating R1300 in revenue—R700 to recover the stock cost and R600 as profit.

This profit is partially reinvested. At $x = 4$, an additional R100 is added to the stock purchase, resulting in a total expenditure of R1400. This includes:

- R700 (original stock cost)
- R600 (previous profit)
- R100 (capital increase)

The capital increase is sourced from the remaining R200 in the account, leaving R100 post-transaction.

Cyclical Dynamics

The model follows a two-month cycle:

- **Month 1**: Capital of R900 is invested to the business.
- **Month 2**: Stock is purchased (negative cash flow).
- **Month 3**: Stock is sold (positive cash flow).
- **Month 4**: Stock is purchased again.
- **Month 5:** Stock is sold again, and so on.

Each cycle increases profit by R600 and requires a capital injection of R100. The alternating sign in the summation reflects this pattern, with odd-indexed months yielding revenue and even-indexed months incurring costs.

Derivatives and Financial Interpretation

First Derivative:

$$\frac{dy}{dx} = (-1)^x(600f(x) + 100g(x) + 700)$$

This expression captures the net cash flow per month:

- Negative values indicate stock purchases and capital injections.
- Positive values represent revenue from stock sales.

Second (Partial) Derivatives:

$\frac{\partial y}{\partial f(x)} = 600$ (Profit rate)

$\frac{\partial y}{\partial g(x)} = 100$ (Capital increase rate)

These derivatives quantify the marginal impact of each transaction type on the account balance.

Example: Profit and Investment After 21 Months
To determine cumulative profit and capital investment at $x = 21$ months, we first derive control functions:

- From the pattern: $x = 2f(x) + 1$

$$f(x) = 10$$

- Similarly: $x = 2g(x) + 3$

$$g(x) = 9$$

Total Profit:

$$\frac{\partial y}{\partial f(x)} \times f(x) = 600 \times 10 = R6000$$

Total Capital Investment:

$$\frac{\partial y}{\partial g(x)} \times g(x) = 100 \times 9 = 900$$

Key Insights

- The model demonstrates how alternating investment and profit cycles affect business liquidity.
- Profit accumulates every second month, while capital injections precede each profit cycle.
- The first derivative reveals the timing of cash inflows and outflows.
- The second derivatives isolate the contribution of profit and investment to the overall balance.

This framework can be adapted to simulate risk, delayed returns, or variable investment rates.

7.2. Application Examples in Science

7.2.1: Application of the Two Common Differences Equation in Fluid Mechanics

We revisit the two-rate linear model introduced in Chapter 1 and apply it to a factory setting involving fluid mixing. Specifically, we examine the process of diluting a liquid chemical with water in a fixed ratio of 6:2—i.e., six litres of chemical to two litres of water. The governing equation is:

$$y = 6f(x) + 2g(x) + 5 \qquad eq.\ 7.2.1.1$$

Variable Definitions

Symbol	Description
(x)	Time in minutes
(f(x))	Volume of chemical dispensed (litres)
(g(x))	Volume of water dispensed (litres)
(y)	Total volume of mixture dispensed (litres)

The constant term 5 represents an initial sample volume already present in the system.

System Description

The dispensing system comprises two input pipes—one for the chemical and one for water—feeding into a bottling station. Bottles are sequentially filled: first with chemical, then with water, adhering to the prescribed ratio. Once filled, each bottle is replaced with an empty one.

The coefficients 6 and 2 in Equation (7.2.1.1) represent the **common differences** or **rates of contribution** for the chemical and water, respectively. These rates can be interpreted as partial derivatives:

$$\frac{\partial y}{\partial f(x)} = 6$$

and

$$\frac{\partial y}{\partial g(x)} = 2$$

These express the marginal increase in total mixture volume per litre of chemical or water added. Since all quantities are measured in litres, the units of these rates are dimensionless (litres/litres), which is consistent and meaningful in this context.

Table 7.2.1.1: Sample Dispensing Data

x	1	2	3	4	5	6	7	8	9
$f(x)$	0	1	1	2	2	3	3	4	4
$g(x)$	0	0	1	1	2	2	3	3	4
y	5	11	13	19	21	27	29	35	37

Control Function Relationships

To determine $f(x)$ and $g(x)$ for arbitrary values of x, we use some of the following control equations:

- For $f(x)$:
    - If x is odd: $x = 2f(x) + 1$
    - If x is even: $x = 2f(x)$
- For $g(x)$:
    - If x is odd: $x = 2g(x) + 1$
    - If x is even: $x = 2g(x) + 2$

These relationships allow us to toggle between time and volume measures, depending on parity.

Example 7.2.1.1: Dispensed Volume at $x = 7$ Minutes

Given $x = 7$ and $g(x) = 3$, we compute:

$$f(x) = \frac{x}{2} - \frac{1}{2} = \frac{7}{2} - \frac{1}{2} = 3$$

And therefore

$$y = 6f(x) + 2g(x) + 5$$

$$y = 6(3) + 2(3) + 5 = 29\ litres$$

Example 7.2.1.2: Dispensed Volume at $x = 80$ Minutes

Since $x = 80$ is even, we use:

$$x = 2f(x)\ and\ x = 2g(x) + 2$$

Therefore

$$f(x) = \frac{1}{2}x = \frac{1}{2}(80) = 40l$$

And

$$g(x) = \frac{1}{2}x - 1 = \frac{1}{2}(80) - 1 = 40 - 1 = 39l$$

$$y = 6(40) + 2(39) + 5 = 240 + 78 + 5 = 323l$$

Thus, after 80 minutes, the system has dispensed:

- 240 litres of chemical
- 78 litres of water
- 5 litres of initial sample

Unit Analysis

Given $f(x) = \frac{1}{2}x$, the coefficient $\frac{1}{2}$ has units of litres per minute. Since x is in minutes, $f(x)$ is in litres. The same applies to $g(x)$. Consequently, the output y is also in litres, and the coefficients 6 and 2 are dimensionless ratios (litres/litres).

Cardinality and Mixing Protocol

To ensure proper mixing and adherence to the 6:2 ratio, we introduce **cardinality constraints** on x. These enforce alternating addition: chemical in one minute, water in the next. This sequencing ensures that the mixture occurs within the bottle and maintains quality control.

Pouring both liquids simultaneously is possible but may violate the ratio unless flow rates are precisely controlled. In practice, alternating positions or nozzle sequencing is preferred to maintain consistency. Mixing is done in bottles but other chemical mixing might require mixing before bottling.

Example 7.2.1.3: Bottle Count at $x = 80$ Minutes

From Example 7.2.1.2:

- Total mixture dispensed: 323 litres
- Initial sample: 5 litres
- Net mixture: (323 - 5 = 318) litres

Chemical contribution: $6 \times 40 = 240$ litres → 40 bottles
Water contribution: $318 - 240 = 78$ litres → 39 bottles

To match the chemical bottle count, an additional 2 litres of water are needed—indicating one bottle was underfilled with water.

7.2.2: Modelling a Three-Liquid Dispensing System Using Control Functions

In this example, we examine a robotic dispensing system designed to fill bottles with three distinct liquids—denoted as liquid 1, liquid 2, and liquid 3. The system operates sequentially, placing and filling bottles one at a time across three designated nozzles. The total volume dispensed over time is modelled by the equation:

$$y = 3f(x) + 3g(x) + 3h(x)$$

Variable Definitions

Symbol	Description
x	Time in minutes
$f(x)$	Quantity of liquid 1 dispensed at time x
$g(x)$	Quantity of liquid 2 dispensed at time x
$h(x)$	Quantity of liquid 3 dispensed at time x
y	Total volume of liquid dispensed (litres)

The coefficients (3) represent the fixed bottle capacity in litres for each liquid. Each control function tracks the number of bottles filled with its respective liquid over time.

System Description

The robotic lever places bottles sequentially into the dispenser's three nozzles. Upon placement, filling begins immediately. The process unfolds as follows:

- **At $x = 2$**: The first bottle is placed and begins filling with liquid 1.
- **At $x = 3$**: The second bottle is placed and begins filling with liquid 2; the first bottle is now one-third full.
- **At $x = 4$**: The third bottle is placed and begins filling with liquid 3; the first bottle is two-thirds full, the second is one-third full.
- **At $x = 5$**: The first bottle is full and removed; the second is two-thirds full, and the third is one-third full.

This staggered placement and filling sequence continues cyclically, governed by the robot's constraint of placing one bottle at a time.

x	1	2	3	4	5	6	7	8	9
$f(x)$	0	1	1	1	2	2	2	3	3
$g(x)$	0	0	1	1	1	2	2	2	3
$h(x)$	0	0	0	1	1	1	2	2	2
y	0	3	6	9	12	15	18	21	24

Table 7.2.2.1: Dispensing Progression

Cardinality and Control Function Relationships

The cardinality of x is 3, reflecting the robot's sequential placement of bottles across three nozzles. This cardinality governs the repetition of control function values and ensures sufficient time for placement, filling, removal, and replacement.

To relate x to each control function, we derive three equations for $f(x)$, based on observed patterns (see Table 7.2.2.1):

- **Equation 1** (for $x = 1, 4, 7, ...$):
$$x = 3f(x) - 2$$
- **Equation 2** (for ($x = 2, 5, 8, ...$):
$$x = 3f(x) - 1$$
- **Equation 3** (for ($x = 3, 6, 9, ...$):
$$x = 3f(x)$$

Similar equations can be derived for $g(x)$ and $h(x)$ using corresponding sequences.

Example 7.2.2.1: Compute y at $x = 8$ Minutes

Given the equation $y = 3f(x) + 3g(x) + 3h(x)$, we determine the control function values:

- From $x = 3f(x) - 1$:
  $8 = 3f(x) - 1$ which means $f(x) = 3$
- From $x = 3g(x) + 2$:
  $8 = 3g(x) + 2$ which means $g(x) = 2$
- From $x = 3h(x) + 2$:
  $8 = 3h(x) + 2$ therefore $h(x) = 2$

Substituting into the main equation:

$$y = 3(3) + 3(2) + 3(2)$$

$$= 9 + 6 + 6 = 21 \; litres$$

This corresponds with the value in Table 7.2.2.1, confirming that after 8 minutes, the system has dispensed:

- 3 bottles of liquid 1
- 2 bottles of liquid 2
- 2 bottles of liquid 3

Each bottle holds 3 litres, yielding a total of 21 litres dispensed.

Conceptual Analogy

The dispensing sequence mirrors a **First-In, First-Out (FIFO)** structure, akin to stack management in computer science or train car sequencing. The robot's lever ensures that the first bottle placed is also the first to be removed, maintaining a consistent and orderly flow.

7.2.3. Modelling Motion with Irregular Sequences and Functions

Objectives

- We want to understand how irregular sequences and control functions can model non-uniform motion.
- We will interpret the composite function $y = 5f(x) + g(x) + 1$ in a real-world context.
- Lastly, we need to analyse the role of each component in capturing operational constraints and dynamic behaviour.

Scenario: Robotic Arm in a Packing Facility

Consider a robotic arm tasked with transferring items from a conveyor belt to a packing station. The motion is not smooth or periodic—it depends on item arrival, packing delays, and mechanical constraints. We model the vertical displacement y of the arm over time x using:

$$y = 5f(x) + g(x) + 1$$

Component	Interpretation
$f(x)$	A control function representing discrete lifting commands (e.g., item pickup).
$g(x)$	An irregular sequence modelling micro-adjustments due to item weight or shape.
Constant $(+1)$	Baseline elevation offset to avoid collision with the conveyor surface.

Table 7.2.3.1: Function Breakdown

- $f(x)$ is typically piecewise constant or stepwise, activating only when an item is detected.
- $g(x)$ may be modeled as a bounded, non-periodic sequence (e.g., $g(x) = \sin(\varphi_x)\Delta_x$) where φ_x and Δ_x vary with item properties.

Diagram: Motion Profile over Time

A schematic diagram can show the arm's elevation over time, with discrete jumps from $f(x)$, fine oscillations from $g(x)$, and a constant offset.

Application Notes

- The coefficient 5 amplifies the control signal, reflecting the mechanical scale of lifting.
- The irregularity in $g(x)$ captures real-world unpredictability—fragile items may require slower descent, heavier ones may trigger dampening.
- The model is ideal for systems where motion is governed by both discrete commands and continuous feedback.

Exercise 7.2.3.1

Problem:
Given $f(x) = \chi_{[2,4]}(x) + \chi_{[6,7]}(x)$ and $g(x) = 0.3\sin(2\pi x)$, plot $y(x) = 5f(x) + g(x) + 1$ over $x \in [0, 10]$.
Question:
Describe how the motion profile reflects the timing and variability of item handling.

Solution:

Given:

- $f(x) = \chi_{[2,4]}(x) + \chi_{[6,7]}(x)$
- $g(x) = 0.3\sin(2\pi x)$
- $y(x) = 5f(x) + g(x) + 1$

Interpretation:

- **Baseline**: $y = 1$ when no item is detected.
- **Active intervals**: In [2, 4] and [6, 7], $f(x) = 1$, so $y \approx 6 + g(x)$, peaking near 6.3 and dipping to 5.7.
- **Oscillations**: The sinusoidal $g(x)$ reflects feedback from item variability—weight, shape, or fragility.

This composite behaviour models a system where discrete commands (pickup) are modulated by continuous feedback (adjustments), ideal for real-world automation. Note that the coefficient before $g(x)$ is taken as 1 and can be any number which will scale $g(x)$ up or down depending on the sign of the coefficient.

Motion Profile Diagram

The plot below visualizes the robotic arm's vertical displacement over time:

- **Blue curve**: $y(x) = 5f(x) + g(x) + 1$
- **Dashed line**: Baseline elevation $y = 1$

The arm lifts during intervals [2, 4] and [6, 7], with sinusoidal oscillations from $g(x) = 0.3\sin(2\pi x)$ adding realistic variability. Outside these intervals, the arm rests at baseline height.

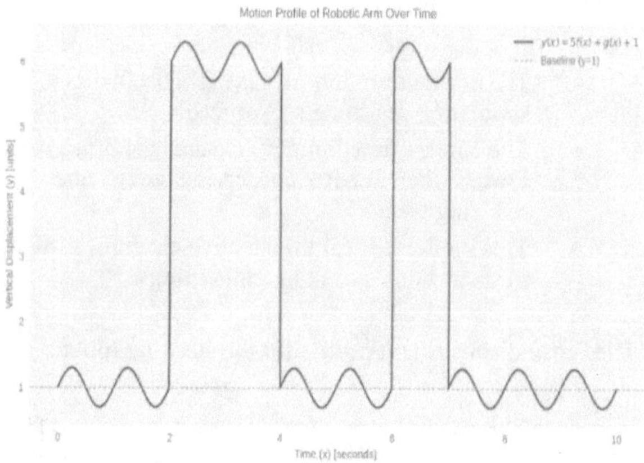

Figure 7.2.3.1

Extension: Service Robotics Example

Scenario: Automated Cleaning Robot

A cleaning robot adjusts its brush height based on surface

type and dirt detection. We model its vertical brush displacement as:

$$y(x) = 4f(x) + 0.5\cos(3\pi x) + 2$$

Component	Interpretation
$f(x)$	Binary signal activating brush lowering when dirt is detected.
$\cos(3\pi x)$	Models surface texture feedback—e.g., tile vs carpet.
Constant (+2)	Keeps brush elevated when inactive.

Table 7.2.3.2

Pedagogical Insight:

- The cosine term introduces rapid oscillations, simulating real-time sensor feedback.
- The control function $f(x)$ ensures the brush only lowers when needed, conserving energy and reducing wear.
- This model is ideal for adaptive cleaning systems in smart homes or industrial settings.

The brush displacement profile for the cleaning robot:

- The brush **lowers** during intervals [1, 3] and [5, 6.5], triggered by dirt detection via the control function $f(x)$.
- The **oscillations** from $0.5\cos(3\pi x)$ simulate real-time surface texture feedback—like transitioning from tile to carpet.
- The brush maintains a **baseline height of 2** when inactive, conserving energy and avoiding unnecessary wear.

Figure 7.2.3.2: Brush displacement profile curve

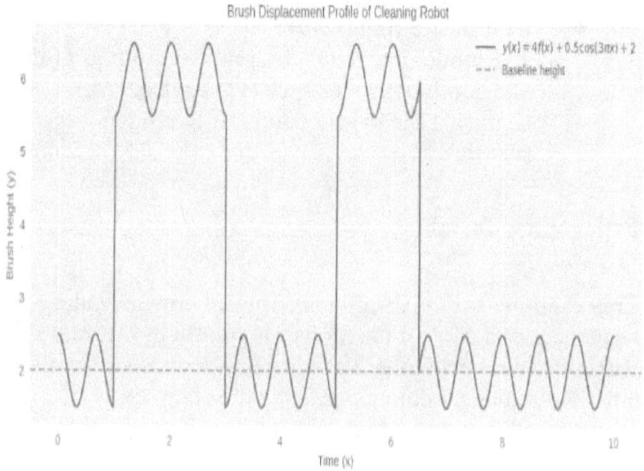

Annotated Diagram: Brush Displacement in Service Robotics

Figure 7.2.3.2 illustrates the vertical displacement profile of a cleaning robot's brush over time, modelled by:

$$y(x) = 4f(x) + 0.5\cos(3\pi x) + 2$$

Key Annotations

- **Baseline Elevation (y = 2):**
  The brush remains elevated when no dirt is detected, preventing unnecessary contact and conserving mechanical energy.
- **Control Activation $f(x) = 1$:**
  During intervals [1, 3] and [5, 6.5], the control function triggers brush descent. The coefficient 4 scales the descent to match operational depth.
- **Oscillatory Feedback $0.5\cos(3\pi x)$:**
  Rapid oscillations simulate real-time surface

texture feedback—e.g., tile vs. carpet. These micro-adjustments reflect adaptive behaviour based on sensor input.

- **Composite Behaviour:**
  The model integrates discrete control (via $f(x)$) and continuous feedback (via cosine term), capturing the hybrid nature of intelligent service robotics.

Pedagogical Insight

This example and its diagram exemplify how irregular sequences and control functions can jointly model real-world robotic behaviour. The blend of stepwise activation and continuous modulation reflects the complexity of modern automation systems, where decisions are both event-driven and sensor-informed.

7.2.4. Irregular sequence control response model compared to PID control response model

Annotated Diagram: Response Curve Comparison

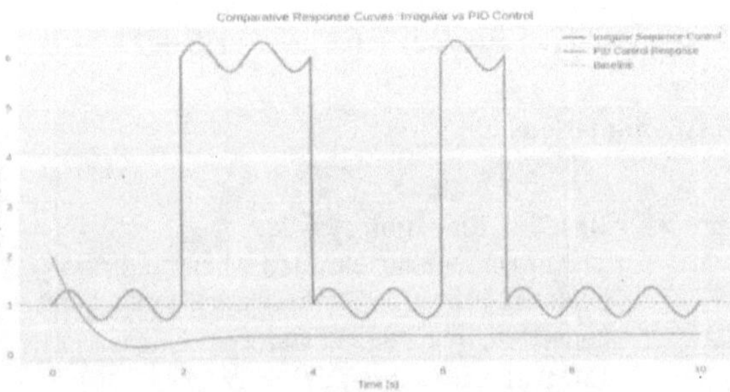

Figure 7.2.4.1: Compares the output profiles of two control paradigms over time:

Curve	Description
Blue Curve	Irregular sequence control: sharp activations during [2, 4] and [6, 7], modulated by sinusoidal feedback. Captures discrete, task-specific behaviour.
Green Curve	PID control response: smooth rise and stabilization with damped oscillations. Reflects continuous error correction typical of second-order systems.
Dashed Line (y = 1)	Baseline or desired set-point for comparison.

Table 7.2.4.1: Comparing output profiles of two control paradigms

---

Insight:

- **Irregular control** is ideal for systems like robotic arms or service bots, where actions are triggered by events (e.g., item detection, dirt presence).
- **PID control** suits systems requiring smooth, continuous adjustment—like temperature regulation or motor speed control.

Pedagogical Exercise: Comparing Control Responses

Problem:
Consider two control models for a robotic brush system:

1. **Irregular Control Model:**
   $y_1(x) = 4f(x) + 0.5\cos(3\pi x) + 2$, where
   $f(x) = \chi_{[1,3]}(x) + \chi_{[5,6.5]}(x)$
2. **PID Control Model:**
   $y_2(t) = K_p e(t) + K_i \int_0^t e(s)ds + K_d(\frac{de(t)}{dt})$,
   where $e(t) = 3 - y_2(t)$

Tasks:

- Plot both responses over $x, t \in [0, 8]$
- Identify intervals where $y_1(x)$ exhibits discrete activation
- Discuss how $y_2(t)$ smooths transitions and fails to reflect activation sparsity

Question:
Which model better captures the operational logic of a brush that only lowers when dirt is detected?

Solution: Given Models

1. Irregular Control Model

$y_1(x) = 4f(x) + 0.5\cos(3\pi x) + 2$, where $f(x) = \chi_{[1,\,3]}(x) + \chi_{[5,\,6.5]}(x)$

- **Discrete activation**: Brush lowers only in intervals [1, 3] and [5, 6.5].
- **Cosine modulation**: Adds realistic variability (e.g., terrain undulation).
- **Baseline**: Outside activation, $y_1(x) = 2 + 0.5\cos(3\pi x)$, oscillating near 2.

2. PID Control Model

$y_2(t) = K_p e(t) + K_i \int_0^t e(s)ds + K_d(\frac{de(t)}{dt})$, where $e(t) = 3 - y_2(t)$

- **Continuous feedback**: Always active, adjusting based on error.
- **Target**: Brush height converges toward 3.
- **Smooth transitions**: No discrete activation logic.

Figure 7.2.4.2: Comparison of control responses for robotic brush system

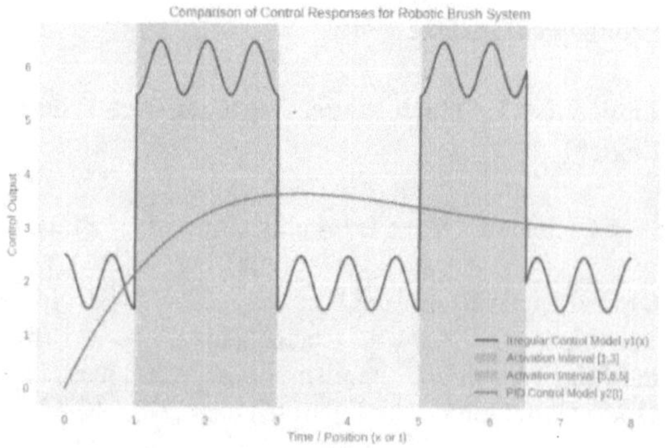

- **Blue plot (Irregular Control)**: Elevated brush height in [1,3] and [5,6.5], with cosine oscillations.
- **Red plot (PID Control)**: Smooth sinusoidal curve, no discrete activation.

Task Breakdown

Discrete Activation Intervals in $y_1(x)$

- **Active**: $x \in [1, 3]$ and $[5, 6.5]$
- **Height**: $y_1(x) \approx 6 \pm 0.5\cos(3\pi x)$
- **Inactive**: Elsewhere, $y_1(x) \approx 2 \pm 0.5\cos(3\pi x)$

PID Response Behaviour

- **No activation sparsity**: Brush is always adjusting.
- **Smooth convergence**: Ideal for maintaining a target height (e.g., 3).

- **Fails to reflect binary logic**: Cannot "turn off" brush when no dirt is present.

Pedagogical Insight

Table 7.2.4.2: Which Model Captures Dirt-Triggered Logic?

Criterion	Irregular Control $y_1(x)$	PID Control $y_2(x)$
Discrete activation	✅ Yes	❌ No
Feedback modulation	✅ Cosine adds realism	✅ Error-based tuning
Operational logic match	✅ Dirt triggers brush lowering	❌ Always active
Sparsity representation	✅ Explicit via (f(x))	❌ Implicit only

Conclusion:
The **Irregular Control Model** better captures the operational logic of a brush that only lowers when dirt is detected. It reflects **event-driven activation**, while the PID model is suited for **continuous regulation**, such as maintaining a fixed brush height regardless of dirt presence. Let's discuss the PID control model a bit, looking specifically to gain coefficients.

Discussing the PID Control Equation

In a PID control system, the parameters K_p, K_i, and K_d are the **gain coefficients** that determine how the controller reacts to error over time.

$$u(t) = K_p e(t) + K_i \int_0^t e(s)ds + K_d(\frac{de(t)}{dt})$$

Where:

- $u(t)$: control output (e.g., motor torque, brush height)
- $e(t)$: error = desired value − actual value

Table 7.2.4.3: Meaning of Each Term

Term	Role in Control	Effect of Increasing Gain
K_p **(Proportional Gain)**	Reacts to **current error**	Faster response, but may overshoot or oscillate
K_i **(Integral Gain)**	Reacts to **accumulated error** over time	Eliminates steady-state error, but can cause instability
K_d **(Derivative Gain)**	Reacts to **rate of change** of error	Dampens oscillations, improves stability, but sensitive to noise

Intuition

- K_p is like saying: "How wrong are we right now?"
- K_i asks: "How long have we been wrong?"
- K_d predicts: "How fast is the error changing?"

Together, they balance **speed, accuracy**, and **stability**.

The PID control model is defined as:

$y_2(t) = K_p e(t) + K_i \int_0^t e(s)ds + K_d(\frac{de(t)}{dt})$, where $e(t) = 3 - y_2(t)$

This creates a **recursive relationship**, since $y_2(t)$ depends on $e(t)$, and $e(t)$ depends on $y_2(t)$. To simulate or plot this system, we typically assume initial conditions and assign values to the gains K_p, K_i and K_d.

Suggested Gain Values for This Problem

To make the comparison pedagogically clear, we can choose:

- $K_p = 1.5$ — moderate proportional response
- $K_i = 0.4$ — slow integral accumulation
- $K_d = 0.2$ — mild damping

These values yield a **smooth, converging response** toward the target value of 3, without excessive oscillation.

The hybrid control model

The **hybrid control model** blends the **event-driven logic** of the irregular control with the **continuous feedback** of PID regulation. This hybrid approach is ideal for systems like a robotic brush that must respond to dirt detection but also adapt to terrain or load variations.

Hybrid Control Model Definition

Let's define the hybrid control output $y_3(x)$ as:

$$y_3(x) \quad is$$

$$= 4f(x) + 0.5\cos(3\pi x) + K_p e(x) + K_i \int_0^x e(s)ds + K_d(\frac{de(x)}{dx}) + 2$$

Where:

- $f(x) = \chi_{[1,3]}(x) + \chi_{[5,6.5]}(x)$
- $e(x) = 3 - y_3(x)$: error from target brush height

And

- $4f(x)$ account for discrete activations
- $0.5\cos(3\pi x)$ – Environmental modulation
- $K_p e(x) + K_i \int_0^x e(s)ds + K_d(\frac{de(x)}{dx})$ – PID feedback

Interpretation

Discrete Activation via $f(x)$

- Brush lowers only when dirt is detected (in intervals [1, 3] and [5, 6.5]).
- Ensures **sparsity** and **energy efficiency**.

Environmental Modulation via $cos(3\pi x)$

- Models terrain undulation or vibration.
- Adds **realistic variability**.

PID Feedback

- Smooth transitions.
- Adapts to **unexpected disturbances** (e.g., uneven surfaces, brush wear).
- Ensures **target height** is maintained during activation.

Table 7.2.4.4: Comparative Summary

Feature	Irregular Control	PID Control	Hybrid Control
Discrete activation	✅	✗	✅
Continuous feedback	✗	✅	✅
Environmental modulation	✅	✗	✅
Stability	Moderate	High	High
Realism	Moderate	Moderate	✅ High
Operational match	Partial	Partial	✅ Full

Pedagogical Insight

The hybrid model:

- **Preserves activation sparsity**: Brush only lowers when needed.
- **Adds adaptive intelligence**: PID terms ensure smooth, responsive behaviour.
- **Models real-world complexity**: Oscillations reflect terrain and load.

Chapter 8

8.1. A Further Investigation

We revisit the class of addition functions characterized by alternating, zigzag behaviour. These functions exhibit a sequence of short, connected linear segments, each possessing its own gradient. As the input variable x increases, the lengths of these segments tend to grow, primarily due to the increasing magnitude of their gradients. This zigzag structure causes the function to oscillate between two bounding straight lines.

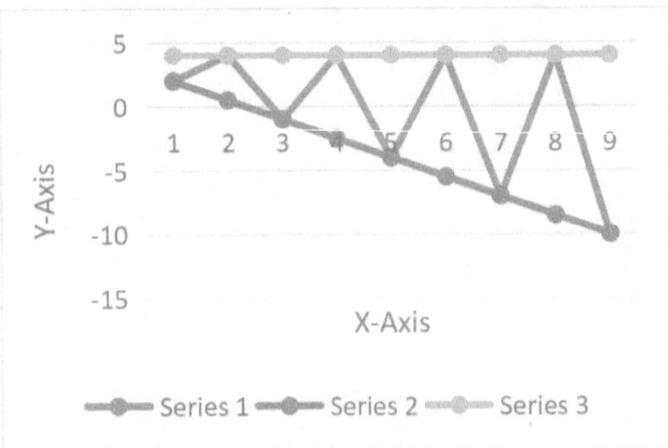

Graph 8.1.1: Illustration of this phenomenon

The points forming the upper and lower edges of the zigzag pattern align along two distinct linear paths. This bracketing behaviour is evident in the tabulated values below.

Table 8.1.1: Zigzag Addition Function

x	1	2	3	4	5	6	7	8	9
$f(x)$	0	1	1	2	2	3	3	4	4
S_x	0	2	-3	2	-6	2	-9	2	-12
y	2	4	-1	4	-4	4	-7	4	-10

The output values y form two interleaved sequences:

- On **odd** x-values: $2, -1, -4, -7, \ldots$
- On **even** x-values: a constant sequence $4, 4, 4, 4, \ldots$

By isolating the constant sequence and analysing the remaining values, we observe that the odd-indexed points lie along a line with gradient $-\frac{3}{2}$. For instance, interpolating between $x = 1$ and $x = 3$, we find intermediate values such as $y = 0.5$ at $x = 2$, and $y =- 2.5$ at $x = 4$, once the constant 4 is removed.

This leads to the identification of two bounding lines:

- **Lower bound**: $y =-\frac{3}{2}x + \frac{7}{2}$
- **Upper bound**: $y = 4$

The gradient $-\frac{3}{2}$ is derived by scaling the alternating difference pattern (i.e., -3) by a factor of (0.5), reflecting the midpoint behaviour between zigzag peaks. The -3 is obtain by taking the common difference of the y values $(2, -1, -4, -7, \ldots)$. The intercept $\frac{7}{2}$ ensures the line passes through the initial point $y = 2$ at $x = 1$, satisfying the condition:

$$2 = -\frac{3}{2}(1) + \frac{7}{2}$$

It is important to note that a naive assumption of the bounding line as $y = -3x + 5$ would misrepresent the structure, failing to account for the even-indexed constant values and the alternating nature of the function.

The main function under investigation is defined as:

$$y = 2 + \sum_{i=1}^{x-1} (-1)^{i-1}(3f(i) + 2)$$

This formulation captures the alternating contributions of the zigzag pattern, weighted by a linear transformation of $f(i)$.

Graphical Interpretation

- **Graph 8.1.1:** correctly illustrates the zigzag function bounded by the two lines.

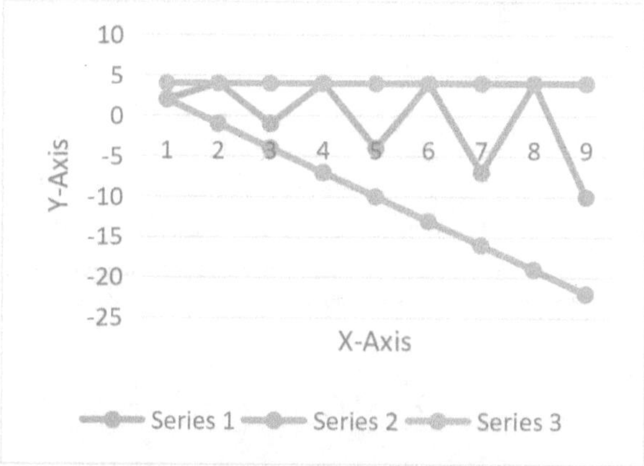

Graph 8.1.2: Based on the incorrect assumption $y =-3x + 5$, fails to encapsulate the bracketing behaviour and misaligns with the observed data.

Boxed Insight: Zigzag Function Behaviour

- The function alternates between two linear bounds, forming a zigzag pattern.
- Each segment has a distinct gradient, contributing to the overall oscillation.
- Proper bounding requires separating constant sequences and analysing midpoint behaviour.
- The correct bounding lines are $y =-\frac{3}{2}x + \frac{7}{2}$ and $y = 4$.

8.2. Function Behaviour between Perpendicular Lines

We consider a function derived from the interaction of two perpendicular lines:

- $y = -2x + 1$
- $y_1 = \frac{1}{2}x + 1$

From these, we construct a sequence of paired values: $-1, \frac{3}{2}, -3, 2, -5, \frac{5}{2}, \ldots$ For instance, at $x = 1$, we obtain $(y, y_1) = (-1, \frac{3}{2})$; at $x = 2$, $(y, y_1) = (-3, 2)$, and so forth. These alternating values define a zigzag sequence governed by the following summation-based function:

$$y = -1 + \sum_{i=1}^{x-1} (-1)^{i-1} \left(\frac{4f(i) + g(i) + 5}{2}\right)$$

To illustrate, we compute:

- For $x = 5$:
  $y = -1 + [\frac{5}{2} + \left(-\frac{9}{2}\right) + 5 - 7] = -5$
- For $x = 6$:
  $y_1 = -1 + [\frac{5}{2} + \left(-\frac{9}{2}\right) + 5 + (-7) + \frac{15}{2}] = \frac{5}{2}$

These results align with the fifth and sixth terms of the sequence, confirming the validity of the function.

Table 8.2.1: Control Function Values

i	1	2	3	4	5	6	7
$f(i)$	0	1	1	2	2	3	3
$g(i)$	0	0	1	1	2	2	3

Graphical Interpretation

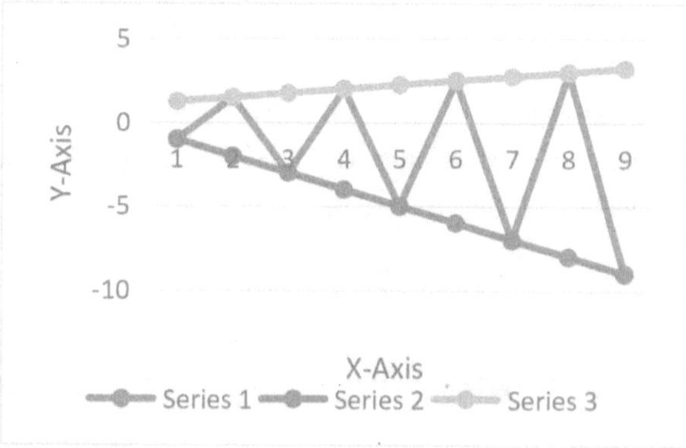

Graph 8.2.1: Bounding the Function

The graph above (i.e. Graph 8.2.1) reveals that the function is bounded between two lines:

- **Lower bound**: $y = -x$ (bottom line, Series2)
- **Upper bound**: $y = \frac{1}{4}x + 1$ (top line, Series3)

To derive these bounding equations, we analyse the behaviour of the function at odd and even x-values.

Odd-indexed y-values:

These form the sequence $-1, -3, -5, -7, \ldots$, which exhibits a constant difference of -2. To interpolate values at even x-positions (e.g., between -1 and -3, we insert intermediate terms such as $-2, -4$, etc., forming a linear sequence with gradient -1. This yields the bounding equation:

$$y = -x$$

Even-indexed y_1-values:

From the sequence $\frac{3}{2}$, 2, $\frac{5}{2}$, ..., we compute the difference between consecutive terms (e.g., $2 - \frac{3}{2} = \frac{1}{2}$, and divide by 2 to obtain the gradient:

$$\frac{1}{2} \div 2 = \frac{1}{4}$$

This leads to the bounding equation:

$$y = \frac{1}{4}x + 1$$

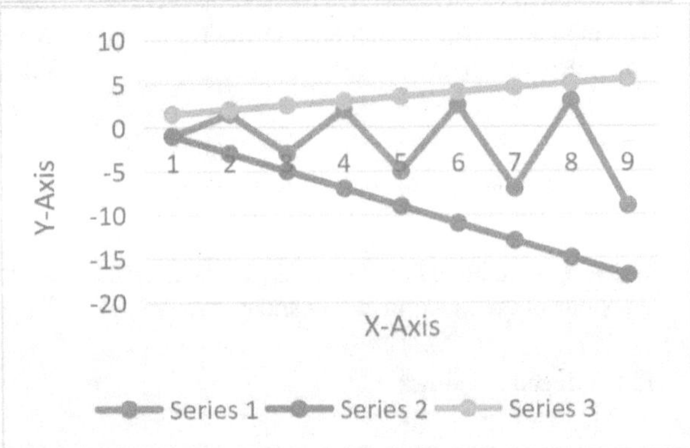

Graph 8.2.2: The graph that depicts the need for adjustment

The Need for Adjustment
If one were to graph the original perpendicular lines $y = -2x + 1$ and $y = \frac{1}{2}x + 1$, as shown in Graph 8.2.2, the resulting bounds would fail to encapsulate the zigzag behavior of the main function. The adjusted bounding lines derived above are necessary to accurately reflect the function's oscillatory structure.

Further discussion of this sequence appears in Part B.3.1 of *Mbele (2024, pp. 13–15)*, where additional properties are explored, though the graph itself was not included.

Boxed Insight: Deriving Bounding Gradients

To determine the gradient of a bounding line from a sequence:

1. **Subtract** the first y-value from the second in the separated sequence.
2. **Divide** the result by 2 (or multiply by 0.5).
3. The outcome is the gradient of the bounding line.

8.3. Comparing Regular and Irregular Linear Functions

We examine two distinct function constructions:

- **Regular linear function**:
  $w = x + 5y + 3$ where $y = x - 1$ (implicitly, since y increments by 1 per unit x).
- **Irregular linear function**:
  $y = f(x) + 5g(x) + 3$ where $f(x)$ and $g(x)$ are staircase functions with periodic repetition.

x	**1**	**2**	**3**	**4**	**5**	**6**	**7**	**8**	**9**
y	0	1	2	3	4	5	6	7	8
w	4	10	16	22	28	34	40	46	52

Table 8.3.1: Regular Linear Function

x	**1**	**2**	**3**	**4**	**5**	**6**	**7**	**8**	**9**
$f(x)$	0	1	1	2	2	3	3	4	4
$g(x)$	0	0	1	1	2	2	3	3	4
y	3	4	9	10	15	16	21	22	27

Table 8.3.2: Irregular Linear Function

Observations and Analysis

1. Rate of Growth

- The regular function w increases by a constant increment of **6**: $\Delta w = w_{x+1} - w_x = 6$
- The irregular function y increases in alternating steps of **1 and 5**, due to the staggered growth of $f(x)$ and $g(x)$. Though the total increment over two steps is also 6, the growth is **non-uniform**.

2. Numerical Structure

- The regular function produces **even values only**, due to its constant linear structure.
- The irregular function yields a **mix of even and odd values**, reflecting its composite nature.

3. Functional Composition

- The regular function is **single-gradient**, defined by a uniform slope.
- The irregular function is **multi-gradient**, composed of two control functions $f(x)$ and $g(x)$, each contributing incrementally.

Boxed Insight: Generalization Potential

Irregular functions possess greater flexibility in modelling diverse behaviours. They can emulate regular functions through appropriate control sequences, but regular functions struggle to replicate irregular patterns without added complexity.

This asymmetry implies that **irregular functions are more generalizable**, capable of encoding both uniform and non-uniform growth, and adapting to layered operational constraints.

Pedagogical Note

The dual-increment structure of the irregular function—where $f(x)$ and $g(x)$ alternate in their contributions—mirrors real-world systems with layered dependencies (e.g., service queues, robotic motion, or commercial transactions). This makes them ideal candidates for applied modelling frameworks.

8.4. Constructing the Ninth Line Segment

To determine the equation of the ninth line segment in the sequence, we begin by identifying two consecutive points through which the line passes. From **Table 8.4.1**, we extract the point $(x, y) = (9, 4)$.

Table 8.4.1: Function values

x	1	2	3	4	5	6	7	8	9
S_x	0	-2	1	-3	2	-4	3	-5	4
y	0	-2	1	-3	2	-4	3	-5	4

To find the next point, corresponding to $x = 10$, we evaluate the summation-defined function:

$$y = \sum_{i=1}^{x-1} (-1)^i (i + 1)$$

Substituting $x = 10$, we compute:

$$= \sum_{i=1}^{9} (-1)^i (i + 1)$$

$$= -2 + 3 - 4 + 5 - 6 + 7 - 8 + 9 - 10$$

$$= -6$$

Thus, the second point is $(10, -6)$. These two points—$(9, 4)$ and $(10, -6)$—define the desired line segment if joint.

Gradient Calculation

The gradient (m) of the line can be determined in two equivalent ways:

Method 1: Using the Summation Derivative

We consider the derivative of the summation function:

$$\frac{dy}{dx} = (-1)^i(i+1)$$

For the ninth line $i = 9$:

$$\frac{dy}{dx} = (-1)^9(9+1) =- 10 = m$$

Method 2: Using Two Points

Alternatively, using the slope formula:

$$m = \frac{y_2 - y_1}{x_2 - x_1} = \frac{-6-4}{10-9} = \frac{-10}{1} =- 10$$

Equation of the Line

Using the point-slope form with point $(9, 4)$ and gradient $m =- 10$:

$$y - 4 =- 10(x - 9)$$

$$y =- 10x + 90 + 4$$

$$y =- 10x + 94$$

Alternatively, using point $(10, -6)$:

$$y + 6 = -10(x - 10)$$

$$y = -10x + 100 - 6$$

$$= -10x + 94$$

Both approaches yield the same linear equation:

$$y = -10x + 94$$

Pedagogical Note

This method illustrates how summation-defined functions can be used to construct piecewise linear segments with alternating gradients. The derivative-based approach offers a direct link between the summation structure and the slope of each segment, while the point-based method reinforces geometric intuition.

Conclusion

Just as the title of the book has reiterated, we have looked at irregular sequences and functions in a way to bring them to the fore so that they are workable as their regular counterpart. We have employed the control function and the cardinality to open analysis to their structure and have them workable.

We have seen a group of two or more straight lines that form a function that depicts a different trend than the ones we are used to, e.g. forming a zig-zag shape. We have also worked with chain functions or hybrid functions which display a unity of different functions as one. The blend of functions into these hybrid functions could be of different straight lines only, straight lines and curved functions and/or different curved functions.

The addition interchanges process has continued to lay more explanative note to some of the discussed sequences or functions. It has done so in conjunction with the summation of common differences, where the later played part in summation functions. The inclusion of what could be an eye-opener in applications of the discussed irregular sequences and functions has continued to refresh us further to look beyond their mathematical structure.

This book initiates a formal exploration of irregular patterns and functions, aiming to render them as analytically tractable as their regular counterparts. While the work presented here is not intended to be definitive, it marks the beginning of a structured investigation into the nature and behaviour of irregularities in mathematical sequences and functions. A central objective is to understand what qualifies these patterns as "irregular" and to delineate their distinctions from regular forms.

References

1. [@heathermagnus-peelvirtuals5935]. (2020, April 8). 3.6. *Transforming irregular functions* [Video]. Heather Magnus-Peel Virtual Secondary School. www.youtube.com/watch?v=9Jb3p33euOM
2. V. (2012). *Everything Maths Grade 12 Mathematics* (p. 5, 15). Siyavula. www.siyavula.com/book-download/maths/Gr12_Mathematics_Learner_Eng_v1.pdf
3. V. (2012). *Everything Maths Grade 10 Mathematics* (p. 293-297). Siyavula. www.siyavula.com/book-download/maths/Gr10_Mathematics_learner_Eng_v1.1.pdf
4. Tsitsiklis, J. N, (2020, December 13). *A few tips on writing papers with mathematical content* [Pdf lecture notes]. https://www.mit.edu/~jnt/papers/R-20-write-v5.pdf
5. Nicholas, J. (2005, June 30). *Sigma notation* [Pdf lecture notes]. Sydney University Press.
6. Steward, J.(2008). *Calculus, early transcendentals* (6th ed.). Thomson Brooks/Cole.
7. Brown, P. (2013). *Sequences and series-A guide for teachers* (year 11-12) (pp. 8-16). Education Services Australia.
8. Anton, H., & Bivens, I., & Davis, S. (2012). *Calculus* (10th ed., pp. 638-639). John Wiley & Sons. Inc..
9. Flenley, R. (2009). *Patterns and algebra* (1st ed., pp. 5-6). 3p Learning Pty (Ltd).
10. Gorts, U. (2023). *Mathematical writing.* Universitat Duisburg Essen.
11. Santos J. C., & Santos M. C. (2015). *Strategies for writing a research paper.* Tourism & Management Studies.

12. [@learnwithmbali]. (2022, April 10). *Irregular patterns* [video]. YouTube. www.youtube.com/watch?v=wD9iF6tNMIA
13. Mynhardt, C. M. (2016). *Introduction to discrete mathematics.* University of South Africa.
14. Haggarty, R. (1992). *Fundamentals of Mathematical Analysis* (2nd ed.). Pearson Education Limited.
15. Mbele, M. E. (2024). A prelude: Introduction to irregular sequences and functions: The addition interchanges process. ScienceOpen, version 2, https://doi.org/10.14293/PR2199.001288.v2
16. Frick, M., & Bohlman, C. A., & Davie, G. & Singleton, J., & Swanepoel, K. J. (2013). *Calculus in Higher Dimensions, Part 1: Basic Concepts.* University of South Africa Press.
17. Pilkington, C. (2017). *Theoretical Computer Science 2: Only Study Guide for COS2601* (pp. 26-28). University of South Africa Press.
18. Borchers, C. (2022). *Grade 12 Mathematics, Past Exam Questions: Equations and Inequalities*. Mr Borchers Maths. https://www.mrborchersmaths.co.za/wp-content/uploads/2022/05/NSC-Gr-12-Past-Exam-Qs-P1-Equations-and-inequalities.pdf
19. Mathcentre, UK. (2009). S*olving inequalities*. Mathcentre. https://www.mathcentre.ac.uk/resources/uploaded/mc-ty-inequalities-2009-1.pdf
20. Thomas, Jr, G. B. (2014). *THOMAS' CALCULUS: EARLY TRANSCENDENTALS* (13TH ED.). Pearson Education, Inc…
21. Mbele, M. E. (2024). Introduction to irregular sequences and functions. ScienceOpen, version 2, https://doi.org/10.14293/PR2199.001317.v2
22. Khan, S. A. (2016, December 16). *Differentiating Power Series*. Khan Academy. Retrieved April 3, 2025, from www.khanacademy.org/math/ap-

calculus-bc/bc-series-new/bc-10-15/v/differentiating-power-series

23. Weiss, W. A. R. (2008, October 2). *Introduction to set theory* [Lecture notes]. https://www.math.toronto.edu/Set_Theory.pdf.
24. Mbele, M. E. (2025). Irregular functions: An introduction. ScienceOpen, version 2, https://doi.org/10.14293/PR2199.001625.v2
25. Lipschutz, S. (1998). *Theory and Problems of Set Theory and Related Topics* (2nd ed.). Mcgraw-hill.
26. University of Houston, UH. (2022). *Basic set theory*. University of Houston. https://www.math.uh.edu/Settheory_Ashlock.pdf.
27. *Basic Concepts of Set Theory: Symbols & Terminology*. (n.d.). Eastern Illinois University. https://castle.eiu.edu/index.
28. Mbele, M. E. (2025). The Extension of the Theory of Straight Lines and Second Degree Irregular Functions. ScienceOpen, version 1, https://doi.org/10.14293/PR2199.002077.v1

URGENT PLEA!

Thank You For Reading My Book!

I really appreciate all of your feedback and
I love hearing what you have to say.

I need your input to make the next version of this
book (and my future books) better.

Please take two minutes now to leave a helpful review
on
Amazon or Draft2Digital letting me know what you
thought of the book.

Thanks so much!

- Mfanufikile Emmanuel Mbele

www.ingramcontent.com/pod-product-compliance
Lightning Source LLC
La Vergne TN
LVHW090319210926
848480LV00001B/297